# Environmental Biocatalysis

# Environmental Biocatalysis: From Remediation to Waste Valorization

Editors

**Evangelos Topakas**
**Jasmina Nikodinovic-Runic**

MDPI • Basel • Beijing • Wuhan • Barcelona • Belgrade • Manchester • Tokyo • Cluj • Tianjin

*Editors*

| | |
|---|---|
| Evangelos Topakas | Jasmina Nikodinovic-Runic |
| School of Chemical Engineering | Institute of Molecular Genetics |
| National Technical University of | and Genetic Engineering |
| Athens | University of Belgrade |
| Athens | Belgrade |
| Greece | Serbia |


*Editorial Office*
MDPI
St. Alban-Anlage 66
4052 Basel, Switzerland

This is a reprint of articles from the Special Issue published online in the open access journal *Catalysts* (ISSN 2073-4344) (available at: www.mdpi.com/journal/catalysts/special_issues/environmental_biocatal).

For citation purposes, cite each article independently as indicated on the article page online and as indicated below:

LastName, A.A.; LastName, B.B.; LastName, C.C. Article Title. *Journal Name* **Year**, *Volume Number*, Page Range.

**ISBN 978-3-0365-1432-1 (Hbk)**
**ISBN 978-3-0365-1431-4 (PDF)**

# Contents

# About the Editors

**Evangelos Topakas**

Evangelos Topakas is an associate professor leading the Industrial Biotechnology and Biocatalysis group at the School of Chemical Engineering of the National Technical University of Athens, Greece. He is the editor-in-chief of the biocatalysis section of the *Catalysts* journal as well as an associate editor for *Frontiers in Bioengineering and Biotechnology* and *Frontiers in Microbiology*. Dr Topakas has also been appointed as a visiting associate professor in the Department of Civil, Environmental, and Natural Resources Engineering, Lulea University of Technology, Sweden, expanding his collaboration to North Europe. His primary objective is to develop new biotechnological tools, enzymes and microorganisms, for the production of second-generation biofuels, bio-based polymers, and chemicals in the concept of industrial biorefinery. The group is active in a broad range of related scientific fields, such as protein engineering, recombinant DNA technology, biochemical engineering, and biocatalysis.

**Jasmina Nikodinovic-Runic**

Jasmina Nikodinovic-Runic is a full research professor, and an Eco-Biotechnology and Drug Development group leader at the Laboratory for Microbial Molecular Genetics and Ecology, Institute of Molecular Genetics and Genetic Engineering, University of Belgrade. She conducts research in the field of molecular genetics of bacteria, directed evolution of enzymes, isolation and characterization of novel biocatalysts, conversion of petrochemical plastic degradation products to biopolymers (PHA) and works on biotechnological process optimizations. Her research interests include microbial biotechnology (bacterial fermentation, strain improvement, and bioremediation), biocatalysis (protein expression and directed evolution, and biotransformations), bacterial bioactive secondary metabolites (biopigments, and antifungal and anticancer compounds), and novel materials (biopolymers, PHA, nanocellulose, and functionalizations). The group is active in eco-green molecular biotechnologies and in the characterization and design of bioactive molecules.

# Preface to "Environmental Biocatalysis: From Remediation to Waste Valorization"

Biocatalysis has developed new molecular tools for the improvement of a wide range of bioprocesses that diminish raw material and energy consumption while reducing or eliminating the formation of byproducts that might be hazardous to human health and the environment. New advances in the field of biocatalysis apply in the environmental sector, where microorganisms or their enzymes could be used for waste valorization towards the production of high added value products or even the mineralization of xenobiotics. Today, such processes have made advanced due to improvements in omics technologies and synthetic biology, which provide an enormous amount of data for genome and metagenome mining, as well as due to the creation of efficient artificial pathways. Novel enzymatic activities may have unique characteristics for the degradation of pollutants as well as high chemo-, regio-, and stereo-selectivity for the modification of aromatic and/or halogenated compounds.

**Evangelos Topakas, Jasmina Nikodinovic-Runic**
*Editors*

catalysts

MDPI

Editorial

# Special Issue on Environmental Biocatalysis

**Jasmina Nikodinovic-Runic [1] and Evangelos Topakas [2,*]**

1 Institute for Molecular Genetics and Genetic Engineering, University of Belgrade, 11000 Belgrade, Serbia; jasmina.nikodinovic@imgge.bg.ac.rs
2 IndBioCat Group, Biotechnology Laboratory, School of Chemical Engineering, National Technical University of Athens, 5 Iroon Polytechniou Str., Zografou Campus, 15780 Athens, Greece
* Correspondence: vtopakas@chemeng.ntua.gr; Tel.: +30-210-772-3264

Received: 24 April 2020; Accepted: 30 April 2020; Published: 1 May 2020

check for updates

Biocatalysis has developed new molecular tools for the improvement of a wide range of bioprocesses that diminish raw material and energy consumption, while reducing or eliminating the formation of byproducts that might be hazardous to human health and the environment. New advances in the field of biocatalysis apply in the environmental sector, where microorganisms or their enzymes could be used for waste valorization towards the production of high added value products or even the mineralization of xenobiotics. Today, such processes have been advanced due to the improvement in omics technologies and synthetic biology, which provide an enormous amount of data for genome and metagenome mining, as well as the creation of efficient artificial pathways. Novel enzymatic activities may have unique characteristics for the degradation of pollutants, as well as high chemo-, regio- and stereo-selectivity for the modification of aromatic and/or halogenated compounds. This Special Issue aims to highlight the dual potential of novel biocatalytic processes, where the first part is dedicated to waste valorization for the production of high value products, while the second part is focused on the detoxification of pollutants [1–8]. Several examples of microbial systems employed for the valorization of waste streams derived by the forest, agricultural and food industries [1–4], or the use of whole-cell or enzyme approaches for the removal of nitrogen or dyes from industrial wastewaters [6–8], were indicated. Last but not least, an example of the utilization of polyhydroxyalkanoates (PHAs) was highlighted for the production of fatty acids, which were used for the enzymatic synthesis of sugar esters with antimicrobial properties [5].

Starting with the use of biocatalysis for the valorization of waste byproducts towards high added value compounds, Karnaouri et al. utilized residual forest biomass for the sustainable production of prebiotic cellobiose and other cellulose-derived oligosaccharides. For this purpose, a fine-tuning of the performance of the commercially available enzyme mixture Celluclast® was conducted towards the optimization of cellobiose production [1]. Under the same idea, Koutrotsios et al. evaluated olive-mill wastes and olive pruning residues as substrates for the cultivation of *Ganoderma lucidum* and *Pleurotus ostreatus*. Both mushrooms exhibited prebiotic properties supporting or even enhancing the growth of both *Lactobacillus acidophilus* and *L. gasseri* bacteria [2]. Lage et al. employed four Nordic green microalgal strains, namely, *Chlorella sorokiniana*, *Chlorella saccharophila*, *Chlorella vulgaris*, and *Coelastrella*, for the valorization of wood hydrolysates (Silver birch, L. *Betula pendula*) and dairy effluent mixture towards lipid production. Culturing microalgae in integrated waste streams under mixotrophic growth regimes was found to be a promising approach for sustainable biofuel production, especially in regions with large seasonal variation in daylight, like northern Sweden [3]. Waste cooking oil, a major pollutant from the food industry, was used as a resource for the high cell density bioprocess for PHA production. The presented bioprocess reached a 33-fold higher PHA productivity compared to previous reports using saponified palm oil, representing an excellent basis for the industrial conversion of waste cooking oil into PHA [4]. In addition to the production of PHA from food waste, Snoch et al. utilized such PHA polyesters for the enzymatic synthesis of sugar esters. With the aid of enzyme biocatalysis, a series of glucose esters were created with unmodified and modified PHA monomers, showing moderate antimicrobial activity [5].

In the second part of the Special Issue, biocatalysis was used as a tool for the treatment of industrial wastewater pollutants. Dolejš et al. studied the co-immobilization of anammox and ammonia-oxidizing bacteria into a polyvinyl alcohol hydrogel, and its effective use in nitrogen removal. This is the first report of such immobilization strategy to indicate that pH-stat and substrate limitation stimulate the co-immobilized bacteria activity in biotransformations [6]. Another form of waste that was considered in this Special Issue was dye-containing wastewater, which presents an increasing global issue due to the increase in population and demand for clothes and other colored products. To tackle this problem, Lončar et al. used genome sequence information to discover dye-decolorizing peroxidases from *Pseudomonas fluorescens* Pf-01, reporting for the first-time the peroxidase-catalyzed insertion of a carbene into an N–H bond [7]. Enzymatic biocatalysis was also employed for the degradation of a broad series of dyes, proving its potential for the treatment of textile industrial wastewater. In particular, Mandic et al. discovered novel laccases combining different approaches including DNA sequence analysis, N-terminal protein sequencing, and genome sequencing data analysis for laccase amplification, cloning, and overexpression. Four active recombinant laccases were obtained from different *Pseudomonas putida* species, highlighting the potential of this genus as a good source of biocatalytically relevant enzymes [8].

Both editors would like to express their gratitude to all the authors for their valuable contributions and the editorial team of Catalysts for their kind support.

## References

1. Karnaouri, A.; Matsakas, L.; Bühler, S.; Muraleedharan, M.; Christakopoulos, P.; Rova, U. Tailoring Celluclast® Cocktail's Performance towards the Production of Prebiotic Cello-Oligosaccharides from Waste Forest Biomass. *Catalysts* **2019**, *9*, 897. [CrossRef]
2. Koutrotsios, G.; Patsou, M.; Mitsou, E.; Bekiaris, G.; Kotsou, M.; Tarantilis, P.; Pletsa, V.; Kyriacou, A.; Zervakis, G. Valorization of Olive By-Products as Substrates for the Cultivation of *Ganoderma lucidum* and *Pleurotus ostreatus* Mushrooms with Enhanced Functional and Prebiotic Properties. *Catalysts* **2019**, *9*, 537. [CrossRef]
3. Lage, S.; Kudahettige, N.; Ferro, L.; Matsakas, L.; Funk, C.; Rova, U.; Gentili, F. Microalgae Cultivation for the Biotransformation of Birch Wood Hydrolysate and Dairy Effluent. *Catalysts* **2019**, *9*, 150. [CrossRef]
4. Ruiz, C.; Kenny, S.; Babu, P.R.; Walsh, M.; Narancic, T.; O'Connor, K. High Cell Density Conversion of Hydrolysed Waste Cooking Oil Fatty Acids into Medium Chain Length Polyhydroxyalkanoate Using *Pseudomonas putida* KT2440. *Catalysts* **2019**, *9*, 468. [CrossRef]
5. Snoch, W.; Stępień, K.; Prajsnar, J.; Staroń, J.; Szaleniec, M.; Guzik, M. Influence of Chemical Modifications of Polyhydroxyalkanoate-Derived Fatty Acids on Their Antimicrobial Properties. *Catalysts* **2019**, *9*, 510. [CrossRef]
6. Dolejš, I.; Stloukal, R.; Rosenberg, M.; Rebroš, M. Nitrogen Removal by Co-Immobilized Anammox and Ammonia-Oxidizing Bacteria in Wastewater Treatment. *Catalysts* **2019**, *9*, 523. [CrossRef]
7. Lončar, N.; Drašković, N.; Božić, N.; Romero, E.; Simić, S.; Opsenica, I.; Vujčić, Z.; Fraaije, M. Expression and Characterization of a Dye-Decolorizing Peroxidase from *Pseudomonas fluorescens* Pf0-1. *Catalysts* **2019**, *9*, 463. [CrossRef]
8. Mandic, M.; Djokic, L.; Nikolaivits, E.; Prodanovic, R.; O'Connor, K.; Jeremic, S.; Topakas, E.; Nikodinovic-Runic, J. Identification and Characterization of New Laccase Biocatalysts from *Pseudomonas* Species Suitable for Degradation of Synthetic Textile Dyes. *Catalysts* **2019**, *9*, 629. [CrossRef]

*catalysts*

Article

# Tailoring Celluclast® Cocktail's Performance towards the Production of Prebiotic Cello-Oligosaccharides from Waste Forest Biomass

**Anthi Karnaouri †, Leonidas Matsakas †, Saskja Bühler, Madhu Nair Muraleedharan, Paul Christakopoulos and Ulrika Rova ***

Biochemical Process Engineering, Division of Chemical Engineering, Department of Civil, Environmental and Natural Resources Engineering, Luleå University of Technology, 971 87 Luleå, Sweden; anthi.karnaouri@ltu.se (A.K.); leonidas.matsakas@ltu.se (L.M.); saskja.buehler@icloud.com (S.B.); madhu.nair.muraleedharan@ltu.se (M.N.M.); paul.christakopoulos@ltu.se (P.C.)

* Correspondence: ulrika.rova@ltu.se; Tel.: +46-0-92-049-1315

† These authors equally contributed to this work.

Received: 22 September 2019; Accepted: 25 October 2019; Published: 28 October 2019

**Abstract:** The main objective of this study focused on the sustainable production of cellobiose and other cellulose-derived oligosaccharides from non-edible sources, more specifically, from forest residues. For this purpose, a fine-tuning of the performance of the commercially available enzyme mixture Celluclast® was conducted towards the optimization of cellobiose production. By enzyme reaction engineering (pH, multi-stage hydrolysis with buffer exchange, addition of β-glucosidase inhibitor), a cellobiose-rich product with a high cellobiose to glucose ratio (37.4) was achieved by utilizing organosolv-pretreated birch biomass. In this way, controlled enzymatic hydrolysis combined with efficient downstream processing, including product recovery and purification through ultrafiltration and nanofiltration, can potentially support the sustainable production of food-grade oligosaccharides from forest biomass. The potential of the hydrolysis product to support the growth of two *Lactobacilli* probiotic strains as a sole carbon source was also demonstrated.

**Keywords:** non-digestible oligosaccharides; Celluclast®; cellobiose; conduritol-B-epoxide; prebiotic; lignocellulose enzyme hydrolysis

---

## 1. Introduction

Nowadays, food and pharmaceutical industries show a growing interest in the development of the so-called functional foods. This term is used to describe products that demonstrate various beneficial effects for the consumer, such as improving the bioavailability of a particular component and, eventually, the reduction of the risk of certain diseases and the general amelioration of the person's well-being [1]. The main target for these high added-value food ingredients are the non- digestible oligosaccharides (NDOs), including carbohydrates with a low degree of polymerization (DP), between 3–10 sugar moieties. Their importance arises from the fact that these oligosaccharides are not digested by the enzymes of the gastrointestinal system, therefore they are available to be fermented by the human intestinal flora and promote the growth of beneficial gut bacteria in the colon, such as *Bifidobacteria* and *Lactobacilli* [2]. Over 20 different types of NDOs are currently available on the market and they are either extracted from natural sources, obtained by enzyme processing, or produced chemically [3]. The preferable process targeted for NDOs production involves the controlled enzymatic degradation, since it offers mild reaction conditions and less chemically harsh byproducts.

A novel source of NDOs are plant cell wall polysaccharides. The inability of humans to digest NDOs derived from cellulose and hemicellulose structures is due to the fact that NDOs sugar units are linked by glucosidic bonds that have a β-configuration and thus cannot be degraded by human gastrointestinal enzymes since they are specific for compounds with an α-configuration [3]. Such plant polysaccharides are often present in large amounts in fiber-rich byproducts and wastes, such as lignocellulosic residues. These residues are currently being underutilized and burned for energy production with a low sale price; this biomass can serve instead as a starting material for many valuable products. There is a rapidly growing interest in new technologies that can convert renewable, low-cost biomass from the forest into high-value bulk chemicals and materials, including NDOs and other food ingredients [4]. The availability of well-defined enzymes or enzyme combinations for the production of NDOs from these substrates is a prerequisite [5]. The plant cell wall polysaccharides as a novel source of prebiotic oligosaccharides have gained increasing attention, since they offer a sustainable and attractive utilization of the agricultural and industrial residues leading to the development of a bio-based economy.

Among the lignocellulosic-derived NDOs, those produced by cellulose hydrolysis, namely cello-oligosaccharides (COS), represent a group of novel oligosaccharides with exceptional interest and numerous potential applications as potential prebiotic ingredients. These oligosaccharides, especially cellobiose, have been shown to promote the growth of *Bifidobacterium* species and exhibit a higher prebiotic potential than other widely used oligosaccharides, such those from fructose [6]. The bioavailability of cellobiose in humans has already been evaluated with cellobiose tolerance tests [7]. It has consequently been observed that after ingestion, cellobiose can be fermented by the gut microflora and that it cannot be hydrolyzed by the digestive enzymes, therefore it reaches the colon undigested [8]. Additional studies have been carried out with humans and rodents and suggested the beneficial effects of cellobiose on carbohydrate metabolism, diabetes and obesity [9,10].

Several strategies for enzyme-mediated production of oligosaccharides have been proposed. These include, among others, the design of tailor-made enzymatic cocktails that offer a controlled polysaccharide cleavage breaking and produce less monomers [11], the modification of reaction conditions (e.g., buffer exchange to abolish the end-product inhibition of enzymatic activity) [12] or the fine-tuning of the performance of commercially available enzyme cocktails (e.g., addition of β-glucosidase inhibitor [13]). Out of these strategies, the construction of customized enzyme mixtures may offer great advantages, such as the ability to adapt the mixture composition to different substrates according to their structural properties, but there are many bottlenecks that hamper the scaling-up of such processes. These bottlenecks arise either from the production costs of different monoenzymes or the laborious, time-consuming techniques related to molecular cloning and heterologous production methods and protocols. On the contrary, tuning the performance of commercially available cellulase mixtures could be a promising solution for the valorization of available biomass wastes. These enzyme cocktails comprise of a unique set of various activities that act synergistically to release the desired product. By modifying the reaction conditions, such as pH or temperature, it is possible to boost the activity of specific enzymatic activities of interest, such as cellobiohydrolases and endoglucanases in the case of cellobiose production, selectively. Combining this selectivity together with blocking the activity of β-glucosidase with inhibitory compounds can theoretically lead to the accumulation of cellobiose. Conduritol-B-epoxide mimics the structure of β-glucose, as shown in Figure 1, and it is a potential irreversible inhibitor of β-glucosidases [14]. Opening of the epoxide by a nucleophilic residue of the enzyme active site enables the interaction with the OH groups of the inhibitor, the formation of a stable ester bond and, subsequently, the specific binding to the enzyme [15].

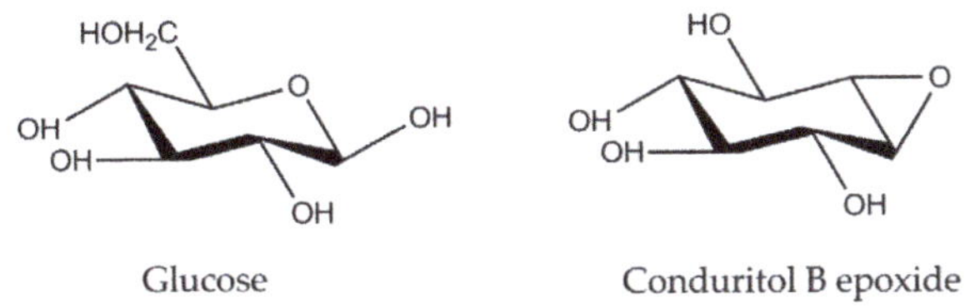


**Figure 1.** Structures of glucose and conduritol-B-epoxide, two potent inhibitors of enzymes with cellobiohydrolase activity.

In the present study, different strategies were followed including modification of the reaction conditions, such as pH, temperature, addition of conduritol-B-epoxide at various concentrations, as well as change of hydrolysis strategies, such as buffer exchange, supplementation of enzyme and conduritol. Celluclast®, an enzyme preparation for breakdown of cellulose into glucose, cellobiose and longer glucose polymers, produced by fungus *Trichoderma reesei*, was selected for the study due to the limited activity of β-glucosidase that it is a native characteristic property of the cocktail [16]. As this enzyme preparation is lacking lytic polysaccharide monooxygenase (LPMO) activities, the synergistic effect of cellulases with external addition of LPMO towards the production of cellobiose was also evaluated. This class of enzymes has been reported to cleave cellulose chains in an oxidative way and to create nicking points [17], thus providing new chain ends for the processive enzymes to act. Moreover, a comparative study of the performance of the optimized cocktail on different substrates under optimal reaction conditions is presented. Organosolv-pretreated birch and spruce woodchips [18,19] were used as substrates to determine the production of cellobiose. Finally, the substrate that was degraded to the greatest extent was assessed on a scale-up reaction and the whole process, including product recovery and purification as well as evaluation of prebiotic potential on different *Lactobacilli* species, is described.

## 2. Results

### 2.1. *Effect of β-Glucosidase Inhibitor*

The effect of conduritol-B-epoxide, which binds covalently to the catalytic site of β-glucosidase, towards the production of cellobiose was evaluated. Organosolv-pretreated birch B1 (the composition is described in Section 4.1), was used as a substrate. Since the inhibition effect is usually dose- and time-dependent, different concentrations of conduritol were tested in order to find the minimum amount that caused inhibition of the activity of β-glucosidase. Buffer exchange was applied at 24 h, as from preliminary results it was found that it increased the hydrolysis yield by 23% (data not shown). Total hydrolysis time was 48 h. As depicted in Figure 2 and Supplementary Table S1, the highest cellulose conversion to cellobiose was observed with a concentration of 0.99 mM conduritol and corresponded to 35.2% (w/w) for 48 h of hydrolysis, while the cellobiose (CB): glucose (Glu) ratio was equal to 3.3. When conduritol was added to a concentration of 3.95 mM, the ratio of cellobiose to glucose was 6.5, with a total production of 168.5 mg cellobiose/g biomass and 25.8 mg Glu/g biomass. In presence of 4.94 mM conduritol, the ratio of CB to Glu was 7.1 and was the highest among all conditions tested, leading to 154 mg CB/g of substrate. In order to combine the maximum ratio of cellobiose to glucose to get a cellobiose-rich hydrolysis product with the minimum amount of conduritol, addition of 3.95 mM conduritol was selected as the optimum condition for further experiments.

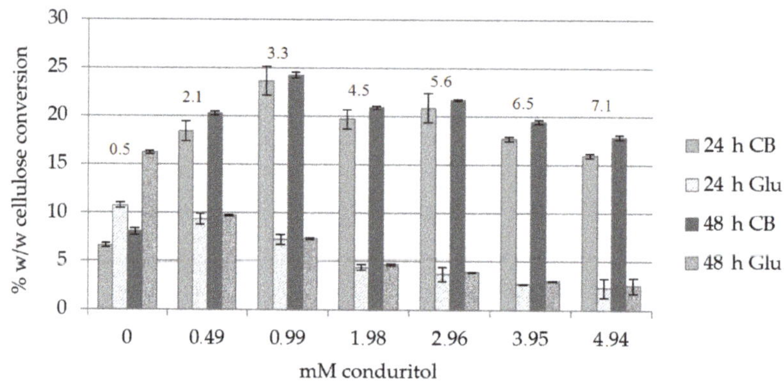


**Figure 2.** Cellulose conversion (% w/w) to cellobiose (CB) and glucose (Glu) under addition of various concentrations of conduritol-B-epoxide, upon buffer exchange at 24 h. The CB:Glu ratio is also described for 48 h of hydrolysis.

### 2.2. *Effect of Enzyme and Conduritol Mixture Preincubation*

Preincubation of enzyme with conduritol was investigated in order to check whether the inhibitor could bind onto the enzyme under a time-dependent mode. The preincubation took place at room temperature for 60 min. Hydrolysis took place under the same conditions as described above and the results were compared to those from a hydrolysis reaction without preincubation. The results, described in Table 1, showed that preincubation did not result in further improvement of the activity of Celluclast® towards the production of CB showing that the covalent bonding between the enzyme and inhibitor is instant. Moreover, the CB:Glu ratio was slightly lower after preincubation.

**Table 1.** Cellulose conversion (% w/w) to CB and Glu with and without preincubation of the enzyme with the inhibitor compound. Birch B1 was used as a substrate.

| | 24 h | | | 48 h | | |
|---|---|---|---|---|---|---|
| | **CB (% w/w)** | **Glu (% w/w)** | **CB:Glu** | **CB (% w/w)** | **Glu (% w/w)** | **CB:Glu** |
| No preincubation | 17.7 ± 0.9 | 2.7 ± 0.1 | 6.7 | 19.5 ± 1.1 | 3.0 ± 1.0 | 6.5 |
| Preincubation | 17.0 ± 1.1 | 2.9 ± 0.0 | 5.8 | 19.9 ± 1.0 | 3.4 ± 1.1 | 5.8 |

### 2.3. *Effect of pH and Enzyme Loading*

The effect of different pH conditions was investigated with an enzyme loading of 25 and 50 mg/g substrate, for 24 and 48 h of hydrolysis, with buffer exchange at 24 h. From the results, as shown in Figure 3 and Supplementary Table S2, it can be concluded that an increase of enzyme loading leads to a concurrent rise of the overall cellulose conversion but it does not favor the production of CB over Glu. It is only at pH 7.0 that the addition of 50 mg enzyme/g of substrate results in production of CB as the sole product in the absence of Glu, with a CB yield of 133 mg/g of substrate after 48 h of hydrolysis confirming previous observations that the inhibition from conduritol-B-epoxide is pH-dependent [20].

The trials at different pH values showed that there is a gradual increase both in total hydrolysis yields and the CB:Glu ratio when the pH of the reaction increases from 4.00 to 6.00, while the optimum CB:Glu ratio is achieved at pH 7.0 (21.6 and 22.5 after 24 and 48 h of hydrolysis, respectively, for 25 mg enzyme/g of substrate). Interestingly, the overall cellulose conversion is lower at this condition, and this is apparently due to the lower activity of a fraction of enzymes that are included in Celluclast® mixture. With the aim to minimize the enzyme usage and, since the amount of CB produced from both enzyme loading conditions was approximately the same, while the amount of glucose was negligible (0.56% w/w), the enzyme loading of 25 mg enzyme/g substrate and pH 7.0 was selected as the optimal condition to proceed further. Trials at pH 7.5 and 8.0 showed a reduced enzyme activity that reached 9.27% and 1.81% w/w cellulose conversion to CB, respectively (data not shown).

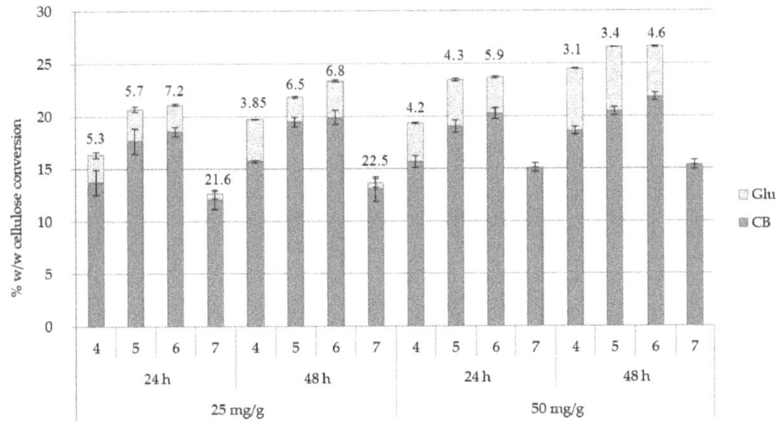


**Figure 3.** Cellulose conversion (% w/w) to cellobiose (CB) and glucose (Glu) in different pH conditions and at an enzyme loading of 25 and 50 mg/g of substrate, upon buffer exchange at 24 h. The CB:Glu ratio is also described for 24 and 48 h of hydrolysis.

## 2.4. *Effect of Hydrolysis Time, Inhibitor Concentration and Reaction Temperature at pH 7.0*

The hydrolysis rates after 24, 48, 72 and 96 h incubation were evaluated at pH 7.00, both with and without buffer exchange at 48 and 72 h, in order to evaluate whether it was possible for the reaction to continue leading to an increase of cellobiose production. Buffer exchange at 24 h was applied in both cases. The results, as depicted in Figure 4a and Supplementary Table S3, showed that the rate is higher during the first 24 h of reaction, with a CB yield of 109.5 mg/g substrate and a CB:Glu ratio of 21.8, which in accordance with the previous results, and it remains high after 48 h. Comparison of 72 and 96 h of hydrolysis with and without buffer exchange did not show significant difference, therefore this strategy was not used further. This can be attributed to the fact that enzymes, though bound onto the substrate, have lost a part of their activity after 96 h of hydrolysis, therefore they are not able to continue the hydrolysis of the substrate and contribute further to the increase of the CB yield. As the greater amount of cellobiose was produced within the first 24 h, this condition was selected in order to minimize the hydrolysis time and, consequently, the possible overall costs of the process.

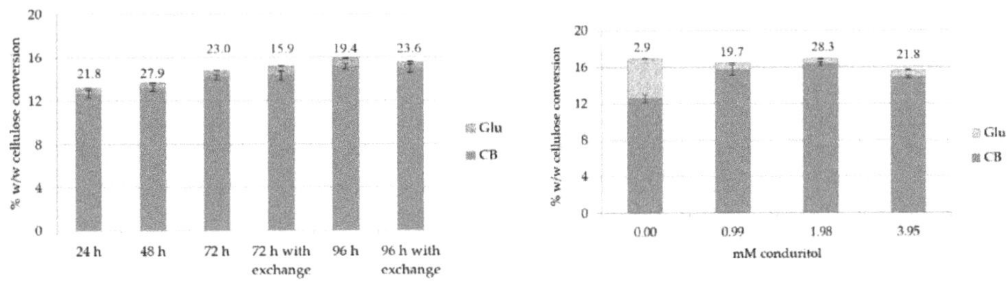


(a) (b)

**Figure 4.** Cellulose conversion (% w/w) to CB and Glu: (**a**) for different hydrolysis time and evaluation of buffer exchange after 72 (buffer exchange at 48 h) and 96 h (buffer exchange at 72 h) of hydrolysis; (**b**) for various concentrations of conduritol-B-epoxide at pH 7.0. The CB:Glu ratio is also described for: (**a**) different hydrolysis time; (**b**) 24 h of reaction.

As pH 7.0 was proved to be a condition that does not favor the activity of β-glucosidase, since the enzyme was apparently less active, and with the aim to reduce the addition of conduritol as much as possible, another set of experiments was set up in order to test different concentrations of the inhibitory compound. Conduritol-B-epoxide was added in concentrations varying within the range of 0–3.95 mM and the reaction was incubated for 24 h. In Figure 4b and Supplementary Table S4, it can be observed that a concentration of 1.98 mM conduritol is sufficient to suppress the activity of β-glucosidase and lead to a total production of 141.7 mg CB/g of substrate, with a CB:Glu ratio of 28.3. Trials with different temperature conditions were also conducted in order to evaluate whether a temperature change within the range of 40–50 °C (optimal temperature conditions for the performance of Celluclast® according to [10]), could have a beneficial effect towards the increase of cellobiose production. The hydrolysis yield and the CB:Glu ratio were evaluated after 12 and 24 h of hydrolysis, after applying buffer exchange at 12 h. The results, summarized in Table 2, showed that 45 °C is the optimal temperature that maximizes the cellobiose production, but the CB:Glu ratio is much lower. Therefore, the condition of 1.98 mM conduritol, 50 °C and 24 h of reaction was chosen as the optimal.

**Table 2.** Trials with different temperature and incubation time. All experiments have been conducted upon the addition of 1.98 mM conduritol-B-epoxide and a pH 7.0, while buffer exchange was applied at 12 h.

| Temperature/Incubation Time | CB | Glu | CB:Glu | mg CB/g Substrate |
|---|---|---|---|---|
| 50 °C, 12 h | 13.0 ± 1.1 | 0.5 ± 0.1 | 27.1 | 111.97 |
| 50 °C, 24 h | 16.3 ± 0.5 | 0.6 ± 0.1 | 28.0 | 140.66 |
| 45 °C, 12 h | 16.7 ± 0.7 | 1.4 ± 0.1 | 12.1 | 143.97 |
| 45 °C, 24 h | 20.1 ± 0.4 | 1.2 ± 0.0 | 17.3 | 173.45 |
| 40 °C, 12 h | 12.8 ± 1.0 | 1.3 ± 0.2 | 9.6 | 110.64 |
| 40 °C, 24 h | 18.5 ± 0.2 | 1.7 ± 0.2 | 10.7 | 159.4 |

### 2.5. *Effect of Buffer Exchange, Enzyme and Inhibitor Supplementation*

The effect of buffer exchange, as well as the supplementation with additional enzyme loading and/or conduritol-B-epoxide was assessed for 8, 24 and 48 h of hydrolysis. Buffer exchange and supplementation of enzyme and inhibitor occurred at 8, 24 and 48 h, while the amount of enzyme and inhibitor added each time was equal to the initial concentration. The results in Figure 5 show that buffer exchange not only improved the CB yield but also increased the CB:Glu ratio at 24 and 48 h. This can be attributed not only to the removal of produced cellobiose and, thus, elimination of end-product inhibition, but also due to the fact that β-glucosidase was removed in the supernatant while other cellulolytic enzymes remained bound onto the substrate. As a result, after the buffer exchange step, the majority of the enzymes that are present represent enzymes with activity of endo- and exo-cellulases, while the β-glucosidase fraction has been removed. Trials with supplementation of enzyme, conduritol or combination of both showed that the addition of enzyme loading boosts the hydrolysis towards the production of cellobiose and rapidly increases the CB:Glu ratio to 39.8, compared to 18.6 in case of buffer exchange. The highest performance is achieved when all three different constituents, namely buffer, enzyme and conduritol, are all supplemented, leading to 164.3 and 172.2 mg CB/g of substrate after 24 and 48 h of hydrolysis, respectively (Supplementary Table S5). Supplementation with conduritol alone does not further improve the product yield.

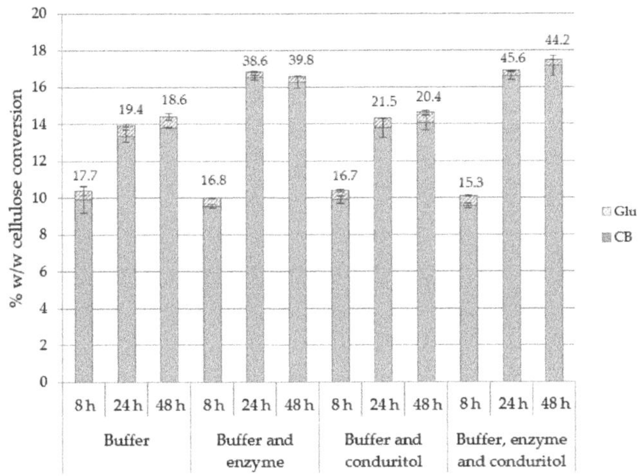


**Figure 5.** Cellulose conversion (% w/w) to CB and Glu after applying buffer exchange and/or supplementation with additional enzyme loading or conduritol-B-epoxide. The CB:Glu ratio is also described for 8, 24 and 48 h of hydrolysis.

## 2.6. Evaluation of LPMO on the Production of Cellobiose

Synergistic effect of *Pc*LPMO9D with cellulases was investigated in order to evaluate whether the addition of a C1 oxidizing enzyme can boost the production of cellobiose. The control reaction with 25 mg Celluclast®/g of substrate and addition of 1.98 mM of conduritol-B-epoxide resulted in the release of 145.7 mg CB/g of substrate, with a CB:Glu ratio equal to 21.8. The test reaction with an additional loading of 2.5 mg LPMO/g of substrate showed a substantial increase of cellobiose to 220.9 g CB/g of substrate. Interestingly, the data displayed in Table 3 show that, in parallel, there is a higher increase in the glucose production rate, which is translated into a decrease of the overall CB:Glu ratio. As a result, although the total sugar release is increased by around 58% in the LPMO supplemented reaction and there is a significant boost obtained in the case of cellobiose, the enormous enhancement in glucose levels causes the CB:Glu ratio to drop to 13.1. Interestingly, in the control reaction containing 27.5 mg Celluclast®/g of substrate, no significant changes in either cellobiose or glucose yields were observed. This is an indication that the increase of cellobiose yields can be attributed to the activity of LPMO and not to the effect of increased protein content. To the best in our knowledge this is the first report where the additional action of LPMO results not only in the increase of release of glucose but also in the increase of production of cellobiose.

**Table 3.** Evaluation of synergistic effect of Celluclast® supplementation with *Pc*LPMO9D towards the production of cellobiose.

| Conditions | CB (% w/w) | Glu (% w/w) | CB:Glu | mg CB/g Substrate | % CB Increase | % Glu Increase |
|---|---|---|---|---|---|---|
| 25 mg enzyme/g sub | 16.0 ± 1.7 | 0.7 ± 0.1 | 21.8 | 145.7 | 0 | 0 |
| 25 mg enzyme/g sub + 2.5 mg *Pc*LPMO9D/g sub | 24.3 ± 1.2 | 1.9 ± 0.1 | 13.1 | 220.9 | 51.5 | 151.6 |
| 27.5 mg enzyme/g sub | 16.0 ± 1.0 | 0.7 ± 1.0 | 21.8 | 145.4 | 0 | 0 |

## 2.7. Evaluation of Different Substrates

Different substrates were tested towards the production of cellobiose by using the optimal hydrolysis conditions, namely 1.98 mM of conduritol-B-epoxide at pH 7.0 with an enzyme loading of 25 mg/g substrate. Buffer exchange was applied at 8 and 24 h of hydrolysis, without any supplementation of conduritol or enzyme. The composition of each substrate is described in Section 4.1. According to the results depicted in Figure 6 and Supplementary Table S6, conversion of cellulose to cellobiose was higher in birch compared to spruce. Regarding the CB:Glu ratio, in the case of birch, this was lower for B1, and in case of spruce it was lower for S1; both were pretreated with acid catalyst. B1 was by far the best substrate among those tested for the production of cellobiose, therefore it was selected for the scale-up reactions.

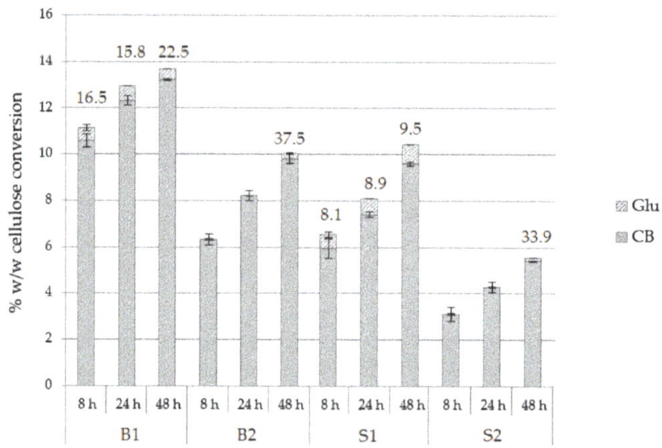


**Figure 6.** Hydrolysis yields from birch and spruce substrates, described as % w/w cellulose conversion into cellobiose and glucose of at pH 7.0, upon the addition of 1.98 mM conduritol-B-epoxide, at an enzyme loading of 25 mg/g of substrate, with buffer exchange at 8 and 24 h. CB:Glu and total mg CB per gram of substrate are also described.

## *2.8. Scale-up Reaction and Downstream Processing for Product Recovery*

A scale-up reaction with a total volume of 60 mL was carried out using birch B1 as the substrate and the optimal reaction conditions to maximize the cellobiose production upon the minimum addition of conduritol-B-epoxide and enzyme loading, as determined in the previous experiments. The hydrolysis was carried out at pH 7.0, with addition of 1.98 mM conduritol-B-epoxide and an enzyme loading of 25 mg/g of substrate. Hydrolysis took place for 24 h and buffer exchange was applied after 8 h. After centrifuge and removal of the residual biomass, all fractions were collected and mixed (final reaction mixture together with that originating from the buffer exchange). Ultrafiltration and nanofiltration were applied in order to remove glucose and conduritol and to recover a cellobiose-rich liquor. Figure 7 represents the overall procedure for the enzymatic production of cellobiose from organosolv-pretreated birch B1, as well as the product yield and recovery in each stage. A total amount of 1.28 g of cellobiose was produced, corresponding to 128 mg of cellobiose/g of substrate, while after ultra- and nano-filtration, 984 mg of the final product remained. The nanofiltration step resulted in the removal of a great amount of glucose, leading to a final cellobiose to glucose ratio of 37.4. The protein content of the mixture was determined at 0.53% w/w. The product was freeze-dried (Figure 8) and used for evaluation of its prebiotic potential.

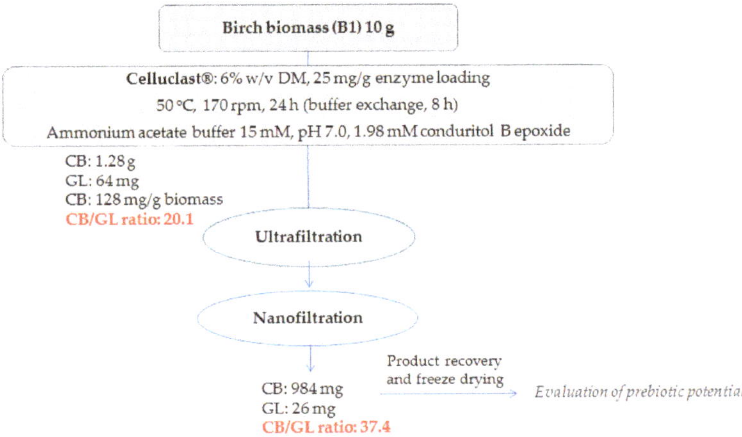


**Figure 7.** Overall scheme of the cellobiose production from pretreated birch.

**Figure 8.** Freeze-dried cellobiose-rich product from enzymatic hydrolysis of pretreated birch.

### 2.9. *Evaluation of COS Prebiotic Activity*

#### Growth Potential of Lactobacillus Strains on Pure Cellobiose and Birch-Derived Sugars

Two *Lactobacilli* strains were used for testing the prebiotic effect of the birch hydrolysates. The results, as evaluated by the increase of the optical density ($OD_{600}$) and the carbohydrate accumulation, are depicted in Figure 9. Both strains (*L. gasseri* and *L. plantarum*) can efficiently utilize cellobiose which is demonstrated by the growth rate values that reach $\mu = 0.19\ h^{-1}$ and $\mu = 0.78\ h^{-1}$, respectively. *L. plantarum* exhibits a relatively higher growth in cellobiose (final $OD_{600}$ = 5.18 ± 0.19) and consumes the total carbohydrate content within the first 23 h of hydrolysis. *L. gasseri* is slower (final $OD_{600}$ = 5.08 ± 0.27) and incubation time over 50 h is required in order to consume the total amount of cellobiose. Lactic acid is the only metabolite that is produced by both *L. gasseri* and *L. plantarum* when grown on cellobiose, as depicted in Table 4, while no production of any short chain fatty acid (acetic, propionic or butyric acid) is detected.

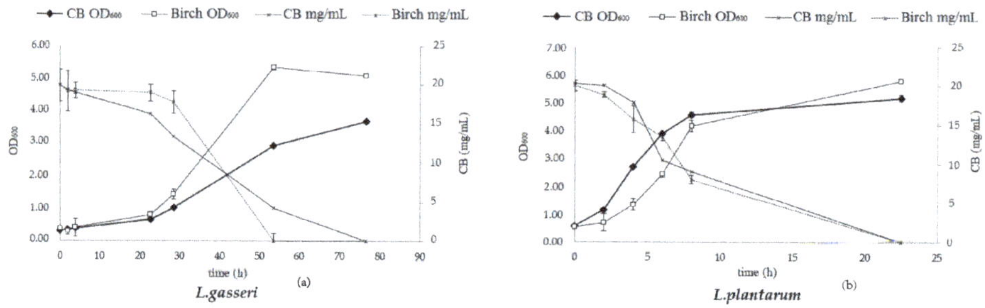


**Figure 9.** Growth curve and carbohydrate consumption of (**a**) *L. gasseri* and (**b**) *L. plantarum* grown anaerobically, at 36 °C, on MRS culture media supplemented with 2% (w/v) cellobiose and with birch hydrolysate at an initial concentration corresponding to 2% (w/v) cellobiose.

**Table 4.** Fermentation metabolites (mg/mL) of *Lactobacilli* strains upon growth on pure cellobiose, birch and spruce COS-rich hydrolysates. No significant amounts of formic or butyric acid were detected.

| | *L. gasseri* | | | | *L. plantarum* | | | |
|---|---|---|---|---|---|---|---|---|
| | **Cellobiose** | | **Birch Hydrolysate** | | **Cellobiose** | | **Birch Hydrolysate** | |
| | **0 h** | **77 h** | **0 h** | **77 h** | **0 h** | **23 h** | **0 h** | **23 h** |
| cellobiose | 19.7 ± 1.1 | 0.1 ± 0.0 | 19.6 ± 2.0 | 0.0 ± 0.00 | 20.1 ± 0.06 | 0.0 ± 0.0 | 20.0 ± 2.6 | 0.1 ± 0.1 |
| lactic acid | 0.1 ± 0.1 | 8.5 ± 1.4 | 0.3 ± 0.1 | 39.5 ± 3.9 | 0.4 ± 0.01 | 21.1 ± 2.9 | 1.2 ± 0.2 | 37.1 ± 2.9 |
| acetic acid | 4.0 ± 1.3 | 4.1 ± 0.0 | 4.2 ± 0.8 | 4.4 ± 1.0 | 4.3 ± 0.25 | 3.8 ± 0.6 | 4.5 ± 1.7 | 4.4 ± 1.9 |
| propionic acid | 0.9 ± 0.0 | 1.03 ± 0.7 | 1.1 ± 0.0 | 0.8 ± 0.2 | 0.9 ± 0.12 | 0.9 ± 0.0 | 1.1 ± 0.7 | 0.8 ± 0.2 |

The results from the studies with birch hydrolysate showed that both strains are able to grow on birch-derived cellobiose as shown in Figure 9 by the increase of the optical density value and the consumption of the cellobiose content. The most effective strain was *L. plantarum*, with a growth rate of $\mu = 0.92\ h^{-1}$, while *L. gasseri* could also utilize this carbon source, though it was much slower ($\mu = 0.22\ h^{-1}$). Determination of the fermentation products (Table 4) reveals the presence of other sugars existing in the biomass hydrolysate that were not detected by HPLC and can be consumed by both strains, since the amount of the lactic acid produced is much higher than the cellobiose that is consumed. In fact, a low amount of glucose is present, but still the final concentration of lactic acid is much higher. Analysis of the product with HPAEC-PAD chromatography revealed traces of sugars with higher degree of polymerization, such as cellotriose and cellotetraose, leading to the conclusion that other oligosaccharides with higher degree of polymerization (DP) originating either for cellulose or hemicellulose exist in the hydrolysate and serve as a carbon source for the bacteria. In a similar way as for the cellobiose substrate, no production of acids was observed.

## 3. Discussion

Lignocellulosic biomass residues constitute an abundant, renewable source of energy-rich polysaccharides that, when subjected to enzymatic hydrolysis, are transformed into oligomeric and monomeric sugars. The latter can either be processed through chemical and/or biological treatment towards a repertoire of high added-value products or, depending on their structural properties, to be used as animal feed or human dietary supplements with prebiotic properties. Many tons of forest materials are annually made available by the forest industry and these residual streams are usually under-valorized. Biomass from birch trees consists of approximately 43.9% w/w cellulose, while this percentage reaches 42% w/w for spruce trees [21], underlining the great potential of these residues towards the production of COS. It has been already reported in the literature that COS produced by enzymatic hydrolysis of wheat straw, comprised of 84% of cellobiose, showed a substantial improvement of the microbial consortium of weaned pigs [22]. However, limited studies on production of COS from forest biomass have yet been reported. Production of non-digestible COS from forest residual biomass includes a complete process starting from physicochemical pretreatment and fractionation in order to obtain a cellulose-rich solid pulp, followed by controlled enzymatic hydrolysis in order to increase the ratio of oligomers over monomers, product recovery and purification and, eventually, evaluation of prebiotic activity. To the best of our knowledge, this is the first study to report the optimization of parameters for tuning the performance of the commercial cocktail Celluclast® towards the production of oligosaccharides and, moreover, this is one of the first reports that describes the whole procedure covering from the pretreatment/fractionation to the in vitro evaluation of COS prebiotic potential on *Lactobacilli* strains.

Lignocellulose degradation is a challenging process due a number of factors linked with the recalcitrance and complex nature of this material. The saccharification of the lignocellulosic biomass is hence limited due to these factors. Particularly, the hemicellulose and lignin removal, the decrease of the degree of polymerization of the cellulose chain and the decrease of the cellulose crystallinity index, which is also accompanied by the decrease of the particle size and thus the increase of the accessible surface area, are considered as major requirements for a hydrolysis process to be efficient [23,24]. Therefore, to boost the hydrolysis of lignocellulose towards the production of prebiotic oligosaccharides, there is a necessity to overcome the difficulty and complexity of the depolymerization of this substrate. Applying an efficient pretreatment/fractionation method, such as organosolv process, has proven a promising solution [18,19], particularly by employing food-grade aqueous ethanol solutions. Combining an efficient fractionation process with controlled enzymatic hydrolysis can boost the release of oligomers over monomers. Synthesis of cellobiose and other oligosaccharides with higher DP is also possible through enzyme-mediated condensation reaction of monosaccharides, catalyzed by β-glucosidases [25,26]. However, the reaction rate is quite slow, leading to very low oligosaccharide yields and thus, making the process not economically viable. Compared to synthesis, hydrolysis of cellulose-containing substrates, including lignocellulosic residues, is an attractive alternative. The main bottleneck of the process is the end-product inhibition, which is more severe in case of cellobiohydrolases [27] but can also affect many enzymatic activities that are present in a cellulase mixture. Additionally, the remaining hemicellulose-derived products on the substrate, like xylan, also impact the hydrolysis rates of cellulases [28]. Even though generally the inhibition is more apparent at high-solid concentrations it is also significant at low-solid concentrations [29] and therefore can occur under the conditions of the present study.

To eliminate the adverse effects of the end-product inhibition, we studied the effect of buffer exchange on the increase of cellobiose yields. A similar strategy described as a multistage degradation process of cellulose towards the production of cellobiose has been also reported in the literature [12]. This process is expected not only to alleviate the inhibitory reactions from present cellobiose and glucose, but also to enable the removal of unbound β-glucosidase in the washing fraction, while other cellulases containing a carbohydrate binding module (CBM) remain attached onto the substrate and continue the hydrolysis process. Moreover, we studied the addition of the β-glucosidase inhibitor conduritol-B-epoxide at different concentrations, which led to a substantial increase of the final cellobiose produced. Finally, modifying the reaction conditions and performing the hydrolysis at pH 7.0 resulted in a high ratio of CB:Glu. This is in accordance with data reported in the literature describing that activity of Celluclast® against pNP-glucopyranoside (indicative of β-glucosidase activity) is observed only in a narrow pH range between 4.0–5.5 [30]. By combining buffer exchange with addition of conduritol and changing the pH of the reaction, we achieved a CB:Glu ratio of 21.8, which is the highest among those reported. Ultrafiltration in order to remove the enzymes and possibly re-use them, combined with nanofiltration in order to remove glucose, conduritol and other low-molecular-weight compounds was also employed for the scale-up reactions [31].

LPMOs, a novel class of oxidative biocatalysts acting on carbohydrate-containing substrates, have been shown to boost the hydrolytic performance of cellulase cocktails by improving their accessibility to the cellulosic substrate [17]. LPMOs have been investigated for their implication in defibrillation and separation of cellulose fibrils, acting as "amorphogenesis"-inducing factors [32], thus providing to hydrolases novel cellulose sites for binding and cleavage. Since LPMOs have this mode of action, it was supposed that they can possibly boost the production of cellobiose. Indeed, as shown in our study, their incorporation in the cellulase mixture increased the yield of CB, but this was accompanied by a simultaneous increase of glucose and, as a result, the decrease of the ratio of CB:Glu. In our case, this condition was not selected further for the experimental purposes of the present project.

Testing different substrates showed that at least two-fold higher cellobiose yield was achieved in the case of birch compared to spruce, verifying that the performance of the cocktail differs among biomass samples and depends on the composition and structure of the lignocellulosic materials that differ according to the type of wood that is used. The main hemicellulolytic component of hardwoods (birch) is xylan, while for softwoods (spruce) is glucomannan. Consequently, the response of the different materials to the pretreatment method that is being used is also distinctive. It has been reported that after the organosolv pretreatment of both birch and spruce biomass the lignin removal is more significant for birch than for spruce and therefore, the enzymatic digestibility showed higher improvement in the case of birch than in the case of spruce [33,34], as it can also be observed by the composition of the OS pretreated substrates that were used for this study.

The results of this study clearly demonstrate that there are several possible strategies that allow a fine-tuning of the performance of the commercial enzyme mixture Celluclast® towards the production of cellobiose. This occurs in a way meaning that there is the option of choosing between high production yields and high purity of the obtained product. Consequently, the decision-making can be adjusted accordingly to the purpose of the experiment. In our case, the target was the production of a cellobiose-rich stream, as pure as possible, in order to be incorporated in products and foods with low glycemic index. This study was a part of ForceUp Value project, funded by Vinnova, Sweden, with the aim to provide functional products. The project's overall approach is to utilize residual lignocellulosic biomass, namely forest feedstocks, for the production of prebiotics to be used as health-beneficial products for human consumption. To achieve this goal towards the production of food supplements with low glycemic index, the focus was on producing COS with a glucose content as low as possible, therefore all experimental design was based on that aspect. The results clearly demonstrate the successful production of COS from birch biomass, as well as their ability to be fermented by beneficial lactic acid bacterial species, which contributes to their prebiotic characteristic.

## 4. Materials and Methods

### 4.1. Enzymes and Substrates

For the production of cellobiose, organosolv-pretreated birch (B1: 200 °C for 30 min, 60% (v/v) EtOH, 1% (w/$w_{biomass}$) $H_2SO_4$; and B2: 200 °C for 15 min, 60% (v/v) EtOH) and spruce (S1: 200 °C for 30 min, 52% (v/v) EtOH, 1% (w/$w_{biomass}$) $H_2SO_4$; and S2: 200 °C for 30 min, 52% (v/v) EtOH) were used as substrates [18,19]. The compositional analysis of the materials was 77.9% (w/w) cellulose, 8.9% (w/w) hemicellulose, 7.0% (w/w) lignin for B1, 66.3% (w/w) cellulose, 22.0% (w/w) hemicellulose, 7.8% (w/w) lignin for B2 and 72.0% (w/w) cellulose, 4.0% (w/w) hemicellulose, 15.4% (w/w) lignin for S1 and 66.0% (w/w) cellulose, 6.0% (w/w) hemicellulose, 14.9% (w/w) lignin for S2 [18,19]. Glucose and cello-oligosaccharides (DP2-6), as well as weak acids (lactic acid, acetic acid, propionic acid) that were used as analytical standards, were obtained from Sigma-Aldrich (St. Louis, MO, USA). Cellulase mixture from *Trichoderma reesei* (Celluclast®) and conduritol-B-epoxide (1,2-anhydro-*myo*-inositol) was purchased from Sigma-Aldrich.

### 4.2. Hydrolysis of Lignocellulosic Materials

Organosolv-pretreated birch B1 was used as a substrate for all experiments towards the optimization of the enzymatic hydrolysis conditions. Organosolv-pretreated birch (B1, B2) and spruce (S1, S2) were then tested as substrates to estimate the cellobiose production from different lignocellulosic feedstocks. All enzymatic treatments took place with a commercially available cellulase mixture from *Trichoderma reesei* (Celluclast®, Sigma-Aldrich). The initial dry matter (DM) in all experiments was 6% (w/v) and the enzyme loading was 25 or 50 mg/g substrate, as described below for each experimental run. Enzymatic reactions were performed in safe lock microtubes at 1.5 mL reaction volume, at 50 °C, under agitation of 1100 rpm. All reactions were performed in 100 mM phosphate-citrate buffer pH 5.0 and contained 0.02% (w/v) $NaN_3$. Conduritol-B-epoxide (Millipore, Burlington, MA, USA) was used as a β-glucosidase inhibitor. Buffer exchange was applied at 24 h, while the total hydrolysis time was 48 h. All trials were run in duplicates. At different time intervals (8, 12, 24, 48 h) according to the design and purpose of each experiment, samples were taken, boiled for 5 min for enzyme inactivation, centrifuged and the supernatant was filtered (0.22 μM pore size). As the main reaction products were cellobiose and glucose, sugar analysis was performed by isocratic ion-exchange chromatography, using an Aminex HPX-87H column (Bio-Rad Laboratories, Hercules, CA, USA) with a micro-guard column at 65 °C as previously described [35]. The % w/w cellulose conversion into cellobiose was calculated by following the equation below:

$$Cellulose\ conversion\ (\%) = \frac{C2 * 1.05 * 100}{Csubstrate * \%\ cellulose * 1.1}, \quad (1)$$

where the concentrations of cellobiose and initial substrate are calculated in mg/mL of reaction volume and 1.05 is the conversion rate of cellobiose to glucose. The % w/w cellulose for each substrate is described in Section 4.1. In case other oligosaccharides were present, the hydrolysates were analyzed with high performance anion exchange chromatography equipped with pulsed amperometric detection (HPAEC-PAD), as previously described [36].

### 4.3. Evaluation of Synergistic Effect of PcLPMO9D with Cellulases towards Cellobiose Production

Combined activity of cellulases with *Pc*LPMO9D from *Phanerochaete chrysosporium* [37] towards the production of cellobiose was evaluated. The enzyme, acting on the C1 atom of the glucose molecules and producing lactones as oxidized products, was heterologously produced in *Pichia pastoris* and purified to its homogeneity according to previously described methods and protocols [37]. The reactions were performed with 6% (w/v) initial DM in 100 mM phosphate-citrate buffer pH 7.0, upon addition of 1.98 mM conduritol-B-epoxide, in the presence of 1 mM ascorbic acid as reducing agent, in safe lock microtubes at 1.5 mL reaction volume, at 50 °C, under agitation of 1100 rpm. Control reaction was performed with 25 mg/g of substrate Celluclast®. To test the effect of *Pc*LPMO9D supplementation, the aforementioned enzyme load was supplemented with 2.5 mg/g of substrate of *Pc*LPMO9D, with the rest of the conditions remaining the same. Another control reaction was also included, in which Celluclast® loading was 27.5 mg/g of substrate. Hydrolysis took place for 8 h. After centrifugation and boiling, the supernatants were filtered (0.22 μM pore size) and the released sugars were detected by HPLC chromatography using an Aminex HPX-87H column as previously described [35].

### 4.4. Scale-up Hydrolysis Reaction and Product Recovery

After identifying the optimal conditions to maximize the cellobiose yield from birch biomass, a scale-up reaction with B1 as a substrate was set up. The initial dry matter was 6% (w/v) and the enzyme loading was 25 mg/g substrate, all suspended in 15 mM ammonium acetate buffer pH 7.0. Ammonium acetate was selected as a sufficiently volatile salt and it was used as a buffer solution at a low concentration in order to minimize the amount of salts in the final product. The hydrolysis total volume was 60 mL in a 500 mL shake flask and the reaction took place at 50 °C, under continuous agitation of 160 rpm, for 24 h. The hydrolysate was collected after centrifugation, filtrated with 0.22 μm pore size filter and then samples were taken for HPLC analysis for identifying and quantifying the cellobiose and glucose content using an Aminex HPX-87H column, as previously described [35]. Then, the hydrolysate was further processed to ultrafiltration for the removal of the enzymes and nanofiltration for the glucose removal and concentration.

For the removal of the total protein, the hydrolysate was filtrated with a LabScale Tangential Flow Filtration system (TFF) (Millipore, Burlington, MA, USA) with exclusion membrane size 5 kDa (Pellicon XL Ultrafiltration Module Biomax 5 kDa, Millipore). The retentate, containing the concentrated solution of cellulases, was maintained in 4 °C for further use in other hydrolysis experiments. The removal of protein was quickly confirmed by determination of protein content with the Bradford method [38]. The permeate was then collected and applied to the nanofiltration system. Nanofiltration was employed for the removal of glucose, conduritol and other low-molecular-weight compounds, as well as for the concentration of birch hydrolysate. A system comprised of a HP4750 High Pressure Stirred Cell (Sterlitech, Kent, WA, USA) and NF270 (pore size ~200–400 Da, Polyamide-TFC, Flux (GFD/psi) 72-98/130, Dow Filmtec™) was employed. A constant pressure of 10 bar was provided by filling nitrogen gas into the cell, while temperature was set at 50 °C. During the nanofiltration process, samples of the permeate and the feed solutions were taken every 15 min on average depending on the flow rate of the permeate. At the end of the filtration, another sample was taken from the sugar mixture that was inside of the vessel (retentate). All samples were then filtrated with 0.22 μm pore size filters and were analyzed with HPLC chromatography using an Aminex HPX-87H column, as previously described in order to determine the sugar content and the presence of acids originating from biomass components (hemicellulose) or reaction conditions (buffer) [35]. The retentate was then collected, freeze-dried and stored in a dry place until further use.

### *4.5. Determination of Prebiotic Potential of Birch Hydrolysate*

Birch hydrolysates produced after enzymatic hydrolysis were tested by prebiotic tests in order to identify whether they could be utilized as carbon sources and support the growth of probiotic strains. *Lactobacillus gasseri* DSM 20077 was purchased from DSMZ (Braunschweig, Germany) while *Lactobacillus plantarum* ATCC 8014 was from ATCC, Manassas, VA, USA. The growth medium for both *Lactobacillus* strains' stock cultures was MRS medium with cysteine (Medium 232 DSMZ). Birch hydrolysate was tested at an initial cellobiose concentration of 2% (w/v) in MRS broth prepared at pH 6.0, in the absence of glucose or any other carbohydrate. The obtained media was then sterilized using 0.22 μm pore size filters. Bacteria cells grown in glucose precultures were centrifuged (4000 rpm, 10 min), collected and resuspended in 50 mL MRS medium containing birch-derived sugars. The cultures were incubated anaerobically, at 36 °C, without agitation, for a maximum of 80 h. Growth rate was monitored by identifying the cell density at 600 nm, while sugar consumption and release of fermentation products (lactic acid, acetic acid, etc.) were analyzed using HPLC chromatography with Aminex HPX-87H column as described above [35]. All trials were run in duplicates. Cultures with MRS media with 2% (w/v) cellobiose were used for comparison.

## 5. Conclusions

The abundance of the lignocellulosic biomass together with its ability to generate high added-value oligosaccharides, such as those derived from the cellulose fraction, through enzymatic treatment make it a sustainable source for the potential larger scale production of these novel food-grade ingredients. In this study, we modified the performance of the commercially available enzyme mixture, Celluclast®, towards the production of COS from birch biomass. The potential of the hydrolysis product to support the growth of two *Lactobacilli* probiotic strains as a sole carbon source was also demonstrated.

**Supplementary Materials:** The following are available online at http://www.mdpi.com/2073-4344/9/11/897/s1, Table S1: Hydrolysis yields after 24 and 48 h of hydrolysis. Cellobiose (CB) and glucose (Glu) production is expressed in % w/w cellulose conversion. Table S2: Effect of pH and enzyme loading on the % w/w cellulose conversion into CB and Glu. Table S3: Cellulose conversion (% w/w) to CB and Glu for different hydrolysis time and evaluation of buffer exchange after 72 and 96 h of hydrolysis. Table S4: Effect of addition of various concentrations of conduritol-B-epoxide at pH 7.0 on the % w/w cellulose conversion into CB and Glu. Table S5: Effect of buffer exchange and/or supplementation with additional enzyme loading or conduritol-B-epoxide on the cellulose conversion (% w/w) to CB and Glu. Table S6: Hydrolysis yields from birch and spruce substrates, described as % w/w cellulose conversion into cellobiose and glucose of at pH 7.0, upon the addition of 1.98 mM conduritol-B-epoxide, at an enzyme loading of 25 mg/g of substrate, with buffer exchange at 8 and 24 h.

**Author Contributions:** Conceptualization, A.K., L.M., U.R. and P.C.; methodology, A.K. and L.M..; investigation, A.K., L.M., S.B., M.N.M.; data curation, A.K. and LM.; writing—original draft preparation, A.K.; writing—review and editing, L.M.; supervision, P.C. and U.R.; project administration, P.C. and U.R.; funding acquisition, U.R. All authors have read and approved the final manuscript.

**Funding:** This work was partially funded by Vinnova, BioInnovation Program Food-grade prebiotic oligosaccharide production, merging marine, and forest resources for moving up the cellulose value-chain (ForceUpValue).

**Acknowledgments:** Eva Grahn Håkansson from Essum Probiotics AB is greatly acknowledged for providing her expertise regarding the prebiotic activity tests. Sveaskog is greatly acknowledged for providing the forest materials. We would like to thank Bing Liu and Mats Sandgren from Swedish University of Agricultural Sciences, Uppsala, for providing the *Pc*LPMO9D. Finally, Bio4Energy, a strategic research environment appointed by the Swedish government, is also acknowledged for supporting this work.

**Conflicts of Interest:** The authors declare no conflict of interest. The funders had no role in the design of the study; in the collection, analyses, or interpretation of data; in the writing of the manuscript, or in the decision to publish the results.

## References

1. Gibson, G.R.; Roberfroid, M.B. Dietary modulation of the human colonic microbiota: Introducing the concept of prebiotics. *J. Nutr.* **1995**, *125*, 1401–1412. [CrossRef] [PubMed]
2. Roberfroid, M. Prebiotics: The concept revisited. *J. Nutr.* **2007**, *137*, 830S–837S. [CrossRef] [PubMed]
3. Swennen, K.; Courtin, C.M.; Delcour, J.A. Non-digestible oligosaccharides with prebiotic properties. *Crit. Rev. Food Sci. Nutr.* **2006**, *46*, 459–471. [CrossRef] [PubMed]
4. Belorkar, S.A.; Gupta, A.K. Oligosaccharides: A boon from nature's desk. *AMB Express* **2016**, *6*, 82. [CrossRef] [PubMed]
5. Mussatto, S.I.; Mancilha, I.M. Non-digestible oligosaccharides: A review. *Carbohydr. Polym.* **2007**, *68*, 587–597. [CrossRef]
6. Pokusaeva, K.; O'Connell-Motherway, M.; Zomer, A.; MacSharry, J.; Fitzgerald, G.F.; van Sinderen, D. Cellodextrin utilization by *Bifidobacterium breve* UCC2003. *Appl. Environ. Microbiol.* **2011**, *77*, 1681–1690. [CrossRef]
7. Nakamura, S.; Oku, T.; Ichinose, M. Bioavailability of cellobiose by tolerance test and breath hydrogen excretion in humans. *Nutrition* **2004**, *20*, 979–983. [CrossRef]
8. Basholli-Salihu, M.; Mueller, M.; Unger, F.M.; Viernstein, H. The use of cellobiose and fructooligosaccharide on growth and stability of *Bifidobacterium infantis* in fermented milk. *Food Nutr. Sci.* **2013**, *4*, 1301. [CrossRef]
9. Satouchi, M.; Watanabe, T.; Wakabayashi, S.; Ohokuma, K.; Koshijima, T.; Kuwahara, M. Digestibility, absorptivity and physiological effects of cellooligosaccharides in human and rat. *Nippon Eiyo Shokuryo Gakkaishi* **1996**, *49*, 143–148. [CrossRef]
10. Watanabe, T. Development of physiological functions of cellooligosaccharides. *Cellul. Commun.* **1998**, *5*, 91–97.
11. Karnaouri, A.; Topakas, E.; Matsakas, L.; Rova, U.; Christakopoulos, P. Fine-tuned enzymatic hydrolysis of organosolv pretreated forest materials for the efficient production of cellobiose. *Front. Chem* **2018**, *6*, 128. [CrossRef] [PubMed]
12. Vanderghem, C.; Boquel, P.; Blecker, C.; Paquot, M. A multistage process to enhance cellobiose production from cellulosic materials. *Appl. Biochem. Biotechnol.* **2010**, *160*, 2300–2307. [CrossRef] [PubMed]
13. Tsuji, A.; Tominaga, K.; Nishiyama, N.; Yuasa, K. Comprehensive enzymatic analysis of the cellulolytic system in digestive fluid of the Sea Hare *Aplysia kurodai*. Efficient glucose release from sea lettuce by synergistic action of 45 kDa endoglucanase and 210 kDa ß-glucosidase. *PLoS ONE* **2013**, *8*, e65418. [CrossRef] [PubMed]
14. Kuo, C.L.; Kallemeijn, W.W.; Lelieveld, L.T.; Mirzaian, M.; Zoutendijk, I.; Vardi, A.; Futerman, A.H.; Meijer, A.H.; Spaink, H.P.; Overkleeft, H.S.; et al. In vivo inactivation of glycosidases by conduritol B epoxide and cyclophellitol as revealed by activity-based protein profiling. *FEBS J.* **2019**, *286*, 3. [CrossRef] [PubMed]
15. Witte, M.D.; van der Marel, G.A.; Aerts, J.M.; Overkleeft, H.S. Irreversible inhibitors and activity-based probes as research tools in chemical glycobiology. *Org. Biomol. Chem.* **2011**, *9*, 5908–5926. [CrossRef]
16. Rodrigues, A.C.; Haven, M.Ø.; Lindedam, J.; Felby, C.; Gama, M. Celluclast and Cellic® CTec2: Saccharification/fermentation of wheat straw, solid-liquid partition and potential of enzyme recycling by alkaline washing. *Enzym. Microb. Technol.* **2015**, *79–80*, 70–77. [CrossRef]

17. Hu, J.; Arantes, V.; Pribowo, A.; Gourlay, K.; Saddler, J. Substrate factors that influence the synergistic interaction of AA9 and cellulases during the enzymatic hydrolysis of biomass. *Energy Environ. Sci.* **2014**, *7*, 2308–2315. [CrossRef]
18. Matsakas, L.; Nitsos, C.; Raghavendran, V.; Yakimenko, O.; Persson, G.; Olsson, E.; Rova, U.; Olsson, L.; Christakopoulos, P. A novel hybrid organosolv: Steam explosion method for the efficient fractionation and pretreatment of birch biomass. *Biotechnol. Biofuels* **2018**, *11*, 1–14. [CrossRef]
19. Matsakas, L.; Raghavendran, V.; Yakimenko, O.; Persson, G.; Olsson, E.; Rova, U.; Olsson, L.; Christakopoulos, P. Lignin-first biomass fractionation using a hybrid organosolv—Steam explosion pretreatment technology improves the saccharification and fermentability of spruce biomass. *Bioresour. Technol.* **2019**, *273*, 521–528. [CrossRef]
20. Umezurike, G.M. The mechanism of action of beta-glucosidase from *Botryodiplodia theobromae* Pat. *Biochem. J.* **1987**, *241*, 455–462. [CrossRef]
21. Willför, S.; Sundberg, A.; Hemming, J.; Holmbom, B. Polysaccharides in some industrially important softwood species. *Wood Sci. Technol.* **2005**, *39*, 245–257. [CrossRef]
22. Jiao, L.F.; Ke, Y.L.; Xiao, K.; Song, Z.H.; Hu, C.H.; Shi, B. Effects of cello-oligosaccharide on intestinal microbiota and epithelial barrier function of weanling pigs. *J. Anim. Sci.* **2015**, *93*, 1157–1164. [CrossRef] [PubMed]
23. Mansfield, S.D.; Mooney, C.; Saddler, J.N. Substrate and enzyme characteristics that limit cellulose hydrolysis. *Biotechnol. Prog.* **1999**, *15*, 804–816. [CrossRef] [PubMed]
24. Zhao, X.; Zhang, L.; Liu, D. Biomass recalcitrance. Part I: The chemical compositions and physical structures affecting the enzymatic hydrolysis of lignocellulose. *Biofuels Bioprod. Biorefin.* **2012**, *6*, 465–482. [CrossRef]
25. Ajisaka, K.; Nishida, H.; Fujimoto, H. The synthesis of oligosaccharides by the reversed hydrolysis reaction of beta-glucosidase at high substrate concentration and at high temperature. *Biotechnol. Lett* **1987**, *9*, 243–248. [CrossRef]
26. Bucke, C. Oligosaccharide synthesis using glycosidases. *J. Chem. Technol. Biotechnol.* **1996**, *67*, 217–220. [CrossRef]
27. Murphy, L.; Bohlin, C.; Baumann, M.J.; Olsen, S.N.; Sørensen, T.H.; Anderson, L.; Borch, K.; Westh, P. Product inhibition of five *Hypocrea jecorina* cellulases. *Enzym. Microb. Technol* **2009**, *52*, 163–169. [CrossRef]
28. Qing, Q.; Yang, B.; Wyman, C.E. Xylooligomers are strong inhibitors of cellulose hydrolysis by enzymes. *Bioresour. Technol.* **2010**, *101*, 9624. [CrossRef]
29. Modenbach, A.A.; Nokes, S.E. Enzymatic hydrolysis of biomass at high-solids loadings—A review. *Biosyst. Agric. Eng. Fac. Publ.* **2013**. [CrossRef]
30. Herlet, J.; Kornberger, P.; Roessler, B.; Glanz, J.; Schwarz, W.H.; Liebl, W.; Zverlov, V.V. A new method to evaluate temperature vs. pH activity profiles for biotechnological relevant enzymes. *Biotechnol. Biofuels* **2017**, *10*, 234. [CrossRef]
31. Qi, B.; Luo, J.; Chen, G.; Chen, X.; Wan, Y. Application of ultrafiltration and nanofiltration for recycling cellulase and concentrating glucose from enzymatic hydrolyzate of steam exploded wheat straw. *Bioresour. Technol.* **2012**, *104*, 466–472. [CrossRef] [PubMed]
32. Villares, A.; Moreau, C.; Bennati-Granier, C.; Garajova, S.; Foucat, L.; Falourd, X.; Saake, B.; Berrin, J.G.; Cathala, B. Lytic polysaccharide monooxygenases disrupt the cellulose fibers structure. *Sci. Rep.* **2017**, *7*, 40262. [CrossRef] [PubMed]
33. Nitsos, C.; Stoklosa, R.; Karnaouri, A.; Vörös, D.; Lange, H.; Hodge, D.; Crestini, C.; Rova, U.; Christakopoulos, P. Isolation and characterization of organosolv and alkaline lignins from hardwood and softwood biomass. *ACS Sustain. Chem. Eng.* **2016**, *4*, 5181–5193. [CrossRef]
34. Raghavendran, V.; Nitsos, C.; Matsakas, L.; Rova, U.; Christakopoulos, P.; Olsson, L. A comparative study of the enzymatic hydrolysis of batch organosolv-pretreated birch and spruce biomass. *AMB Express* **2018**, *8*, 114. [CrossRef]
35. Karnaouri, A.; Rova, U.; Christakopoulos, P. Effect of different pretreatment methods on birch outer bark: New biorefinery routes. *Molecules* **2016**, *21*, 427. [CrossRef] [PubMed]
36. Karnaouri, A.; Muraleedharan, M.N.; Dimarogona, M.; Topakas, E.; Rova, U.; Sandgren, M.; Christakopoulos, P. Recombinant expression of thermostable processive *Mt*EG5 endoglucanase and its synergism with *Mt*LPMO from *Myceliophthora thermophila* during the hydrolysis of lignocellulosic substrates. *Biotechnol. Biofuels* **2017**, *10*, 126. [CrossRef] [PubMed]

37. Westereng, B.; Ishida, T.; Vaaje-Kolstad, G.; Wu, M.; Eijsink, V.G.; Igarashi, K.; Samejima, M.; Ståhlberg, J.; Horn, S.J.; Sandgren, M. The putative endoglucanase *Pc*GH61D from *Phanerochaete chrysosporium* is a metal-dependent oxidative enzyme that cleaves cellulose. *PLoS ONE* **2011**, *6*, e27807. [CrossRef]
38. Bradford, M.M. A rapid and sensitive method for the quantitation of microgram quantities of protein utilizing the principle of protein-dye binding. *Anal. Biochem.* **1976**, *72*, 248–254. [CrossRef]

catalysts

MDPI

*Article*

# Valorization of Olive By-Products as Substrates for the Cultivation of *Ganoderma lucidum* and *Pleurotus ostreatus* Mushrooms with Enhanced Functional and Prebiotic Properties

**Georgios Koutrotsios [1], Marianna Patsou [2], Evdokia K. Mitsou [2], Georgios Bekiaris [1], Maria Kotsou [2], Petros A. Tarantilis [3], Vasiliki Pletsa [4], Adamantini Kyriacou [2] and Georgios I. Zervakis [1,*]**

1 Laboratory of General and Agricultural Microbiology, Department of Crop Science, Agricultural University of Athens, 11855 Athens, Greece; georgioskoutrotsios@gmail.com (G.K.); giorgosbekiaris@yahoo.gr (G.B.)
2 Department of Nutrition and Dietetics, Harokopio University, 17676 Athens, Greece; patsoumarianna@windowslive.com (M.P.); emitsou@hua.gr (E.K.M.); mkotsou@hua.gr (M.K.); kyriacou@hua.gr (A.K.)
3 Laboratory of Chemistry, Department of Food Science & Human Nutrition, Agricultural University of Athens, 11855 Athens, Greece; ptara@aua.gr
4 Institute of Chemical Biology, National Hellenic Research Foundation, 11634 Athens, Greece; vpletsa@eie.gr
* Correspondence: zervakis@aua.gr; Tel.: +30-210-529-4341

Received: 20 May 2019; Accepted: 11 June 2019; Published: 16 June 2019

check for updates

**Abstract:** The successful management of olive by-products constitutes a major challenge due to their huge volume, high organic content, and toxicity. Olive-mill wastes (TPOMW) and olive pruning residues (OLPR) were evaluated as substrates for the cultivation of *Ganoderma lucidum* and *Pleurotus ostreatus*. Chemical composition, glucans, total phenolic content, and antioxidant activity were measured in mushrooms, and their prebiotic potential was assessed by examining their effect on the growth of four intestinal bacteria. Several substrates based on olive by-products had a positive impact on *P. ostreatus* mushroom production, whereas only one performed adequately for *G. lucidum*. Increased ratios of OLPR to wheat-straw resulted in an increase of crude protein content in *P. ostreatus* fruit-bodies by up to 42%, while *G. lucidum* mushrooms from OLPR-based substrates exhibited an up to three-fold increase in α-glucan, or a significant enhancement of β-glucan content, when compared to beech sawdust (control). The mushrooms' FTIR spectra confirmed the qualitative/quantitative differentiation detected by standard assays. In regard to prebiotic properties, mushrooms powder supported or even enhanced growth of both *Lactobacillus acidophilus* and *L. gasseri* after 24/48 h of incubation. In contrast, a strain-specific pattern was observed in bifidobacteria; mushrooms hindered *Bifidobacterium bifidum* growth, whereas they supported a similar-to-glucose growth for *B. longum*.

**Keywords:** olive mill waste; lignocellulosic residues; *Ganoderma lucidum*; *Pleurotus ostreatus*; medicinal mushrooms; glucan; prebiotic; *Lactobacillus*; *Bifidobacterium*; waste valorization

## 1. Introduction

The particular organoleptic and nutritional properties of olive oil, in conjunction with the need to improve human diet, have resulted at a continuous increase in olives production during the last few decades. Nowadays, over 750 million olive trees are cultivated worldwide, 95% of which grow in the Mediterranean region, while the global production of olive oil is projected to reach 3.1 million metric tons in 2018/2019 (International Olive Council, http://www.internationaloliveoil.org). However, the cultivation of olive trees and operation of olive mills generate huge amounts of plant residues and effluents, respectively. The annual production of pruning exceeds 15 million tons, all of which are usually burnt on farms and result in air pollution due to aerosols rich in organic compounds [1]. Furthermore, large quantities of a highly toxic and recalcitrant sludge-like waste (known as "alperujo") are produced by two-phase olive oil mills [2]. Moreover, centralized alperujo treatment is not feasible in most countries due to the olive mills' small capacity, seasonal operation, and scattered distribution [3].

Olive by-products—especially olive-oil mill waste—have attracted scientific interest, and various physicochemical or biological processes have been proposed for reducing their pollution load [4–7]. However, their wide-scale implementation is often technically complicated and/or not financially viable. Hence, the adoption of innovative processes for the generation of value-added products presents an alternative worth investigating in order to successfully exploit such wastes by reducing their environmental impact.

Mushroom cultivation constitutes a noteworthy and sustainable practice through which lignocellulosic residues are enzymatically biotransformed into fungal biomass with particular nutritional and/or medicinal properties. Though cereal straw and hardwood sawdust are commonly used as substrates for the production of most mushrooms species [8], many of them are also cultivated on various residues, e.g., cottonseed hulls, corn cobs, sugarcane bagasse, cotton gin trash, coffee husks, grape marc, vineyard pruning, banana straw, palm leaves, soybean stalk, waste paper, and nut shells [9–17], as well as on olive mill wastes [18–21]. Among the most widely appreciated mushrooms are those produced by *Pleurotus ostreatus* and *Ganoderma lucidum*. The former (commonly known as the 'oyster mushroom') is the third most cultivated species worldwide [22]. Its culinary/nutritional value and the relatively easy cultivation process has resulted in the large spread of its cultivation over the last 20 years. Mushrooms, as well as the mycelia of *P. ostreatus*, contain bioactive compounds such as polysaccharides, lectins, lipopolysaccharides, peptides, and triterpenoids to which a plethora of medicinal properties are attributed, including anticancer, antitumor, anti-inflammatory, immunostimulatory, and immunomodulatory activities [23]. On the other hand, *G. lucidum*, commonly known as the reishi mushroom, has been extensively used as a pharmaceutical product. Over 400 chemical compounds present in *G. lucidum* biomass have been classified as bioactive, including polysaccharides, proteins/peptides, steroids, sterols, and fatty acids, which are associated to antioxidative, antiaging, antifatigue, hypoglycemic, immunomodulating, anti-inflammatory, antitumor, antibacterial, antiviral, hypolipidemic, sleep regulating, and analgesic properties [23,24]. Furthermore, both *P. ostreatus* and *G. lucidum* have demonstrated promising prebiotic properties, possibly due to their indigestible polysaccharides and, particularly, β-glucan content [25,26].

Fungal (mushroom) polysaccharides comprise chitin, α- and β-glucans, mannans, xylans, and galactans; they are mainly found as linear and branched glucans with various glycosidic linkages, e.g., (1→3), (1→6)-β-glucans, and (1→3)-α-glucans, while some contain arabinose, glucuronic acid, galactose, mannose, xylose, or ribose [27]. Glucans, in particular, have deservedly earned much attention thanks to their exceptional but not yet fully understood immunobiological activity [28]. Their presence in several mushroom species has been associated with various functional and medicinal properties [29–33].

The nutritional composition of mushrooms, as well as their content in various bioactive ingredients, are differentiated quantitatively and qualitatively depending on the strain, cultivation conditions, and substrate used, as previously demonstrated with *Cyclocybe cylindracea*, *Hericium erinaceus*, and *Pleurotus* spp. [10,18,19,34,35]. In general, it has been shown that the addition of materials rich in phenolic and antioxidant compounds in cultivation substrates of the aforementioned species lead to a significant increase in the respective components in fruit-bodies. However, no data exist about the cultivation of *G. lucidum* on olive by-products/wastes or in regard to the effect of production substrates on mushroom composition and functional properties. In addition, very limited information is available on the prebiotic properties of lyophilized mushroom powder deriving from different species, cultivation processes, and media [25].

In the frame of the present study, the suitability of olive mill wastes and olive cultivation residues (in comparison to commonly/widely used substrates) for the production of *G. lucidum* and *P. ostreatus* mushrooms was examined. In addition to cultivation parameters, both the crude composition and the content of fruit-bodies in selected bioactive compounds were evaluated, and the effect of various substrates was assessed. Finally, the prebiotic potential of mushrooms was investigated in vitro based on selected intestinal microbial strains cultivation.

## 2. Results and Discussion

### 2.1. Initial Assessment of Substrates for Fungal Growth

The first part of this study aimed at determining the growth of *G. lucidum* and *P. ostreatus* in substrates consisting of olive pruning residues (OLPR) and two-phase olive mill wastes (TPOMW) in various mixtures with each other or with beech sawdust (BS) and wheat straw (WS), respectively. The growth rate of *G. lucidum* was significantly higher on BS as compared to all other substrates (Figure 1). In general, values in 'race tubes' containing up to 50% OLPR were comparable to those measured in 50% TPOMW. Increasing the OLPR content led to significant decrease in growth rates, while mixtures of OLPR and TPOMW were the worst performing media and were excluded from further experiments. In the case of *P. ostreatus*, the control (WS) and the WS:TPOMW 3:1 substrate equally supported mycelium growth; however, growth rates were significantly reduced when TPOMW or OLPR were added at a ratio of 50% or higher (Figure 1). It is worth mentioning that the three ratios of OLPR to WS did not differ significantly, whereas the OLPR substrate alone, as well as the OLPR and TPOMW mixtures, showed relatively low growth rates.

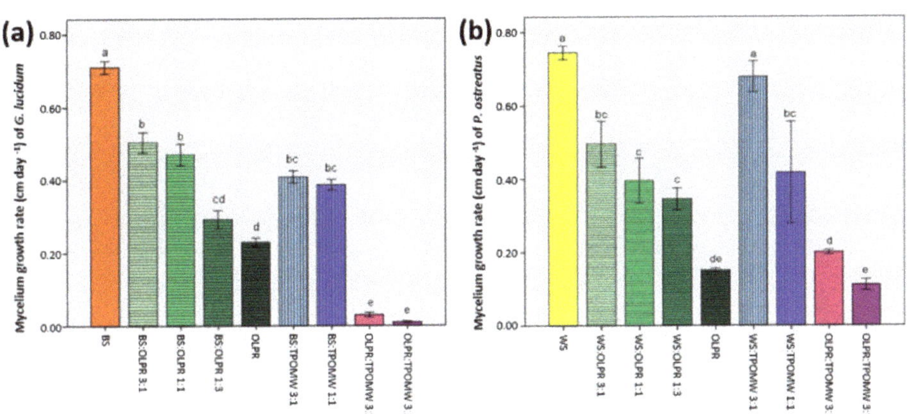


**Figure 1.** Mycelium growth rates of (**a**) *Ganoderma lucidum* and (**b**) *Pleurotus ostreatus* during the colonization of nine substrates prepared on the basis of three main ingredients (BS: Beech sawdust, WS: Wheat straw, TPOMW: Two-phase olive mill waste, and OLPR: Olive pruning residues) and their various mixtures as estimated in "race" tube experiments. Values (cm $day^{-1}$) are expressed as means ± standard errors of means, $n = 4$. A lack of letters in common indicates statistically significant differences (Gabriel's *t*-test, $p < 0.05$) for comparisons of treatment means between different substrates.

Hence, the outcome of this initial assessment revealed that increasing amounts of TPOMW in substrates retarded mycelium growth due to the elevated toxicity exerted mainly by the higher polyphenolic content. This effect was more pronounced for *G. lucidum*, while similar behavior was noted in OLPR-based media for both species examined. Especially in regard to *P. ostreatus*, a reduction of growth was only observed when olive by-products were used at high ratios. Such results are in accordance to earlier findings, where the addition of composted (or not) TPOMW at a ratio of up to 20% in substrates of several *Pleurotus* species (including *P. ostreatus*) contributed to a growth increase, which was adversely affected when TPOMW exceeded 40% [21]. However, the fungal growth response to the nature of substrates also depends on the species examined. Thus, and in contrast to what was observed in the present study for *G. lucidum*, *H. erinaceus* strains showed satisfactory growth in substrates containing OLPR or TPOMW, exceeding, in most cases, the values obtained by the BS control [18].

### 2.2. Evaluation of TPOMW- and OLPR-Based Substrates for Mushroom Cultivation

Previously obtained results led to the elimination of the worst performing substrates. Thus, OLPR 25%, 50%, and 75%, and TPOMW 25% and 50%, in mixtures with the conventional substrates (i.e., BS for *G. lucidum* and WS for *P. ostreatus*; BS or WS alone were also used as controls for each species) were further examined in regard to their suitability to support mushroom production in comparison to the control substrates.

*G. lucidum* completed incubation within 26–33 days in most of the substrates examined, but it performed significantly slower on BS and BS:OLPR 3:1 (up to 43 days; Table 1). In contrast, these two particular substrates (BS and BS:OLPR 3:1) provided the best earliness (time required for the appearance of mushroom primordia) values (45 to 46 days), while the rest were notably slower, in particular those containing TPOMW (77–84 days). Total yield ranged from 19 to 275 g, and the corresponding biological efficiency (BE; i.e., fresh weight of mushrooms produced over the dry weight of the substrate) values varied from 5 to 61% (Table 1). The control substrate (BS) provided significant higher values in respect to the other treatments, while only BS:OLPR 3:1 exhibited a satisfactory performance among all substrates containing olive by-products.

**Table 1.** Mushroom cultivation parameters for *Ganoderma lucidum* and *Pleurotus ostreatus* in six substrates consisting of beech sawdust (BS), wheat straw (WS), two-phase olive mill waste (TPOMW), and olive pruning residues (OLPR) in various mixtures (*w*/*w*). Values are expressed are means ± standard deviation of means, $n = 4$. Lack of superscript letters in common indicates statistically significant differences (Gabriel's *t*-test, $p < 0.05$) for comparisons between substrates.

| | | | *Ganoderma lucidum* | | | |
|---|---|---|---|---|---|---|
| | **BS** | **BS:OLPR 3:1** | **BS:OLPR 1:1** | **BS:OLPR 1:3** | **BS:TPOMW 3:1** | **BS:TPOMW 1:1** |
| Incubation period (days) | $43.00 \pm 0.00^{a}$ | $43.00 \pm 0.00^{a}$ | $26.33 \pm 0.67^{c}$ | $31.25 \pm 0.63^{b}$ | $33.00 \pm 1.16^{b}$ | $26.33 \pm 0.67^{c}$ |
| Earliness (days) | $45.75 \pm 0.25^{b}$ | $45.00 \pm 3.68^{b}$ | $52.75 \pm 4.50^{ab}$ | $63.67 \pm 15.50^{ab}$ | $77.00 \pm 18.54^{a}$ | $83.50 \pm 19.50^{a}$ |
| Total yield (g) | $275.22 \pm 14.48^{a}$ | $193.42 \pm 13.52^{b}$ | $61.46 \pm 10.8^{c}$ | $19.10 \pm 8.3^{d}$ | $36.10 \pm 14.47^{cd}$ | $61.46 \pm 10.84^{c}$ |
| Biological efficiency (%) | $61.24 \pm 3.22^{a}$ | $40.24 \pm 2.81^{b}$ | $20.52 \pm 3.62^{c}$ | $4.54 \pm 1.97^{d}$ | $12.05 \pm 4.83^{c}$ | $20.52 \pm 3.62^{c}$ |
| | | | *Pleurotus ostreatus* | | | |
| | **WS** | **WS:OLPR 3:1** | **WS:OLPR 1:1** | **WS:OLPR 1:3** | **WS:TPOMW 3:1** | **WS:TPOMW 1:1** |
| Incubation period (days) | $26.00 \pm 0.00^{c}$ | $28.00 \pm 0.33^{b}$ | $30.67 \pm 2.33^{ab}$ | $36.47 \pm 2.19^{a}$ | $25.00 \pm 0.00^{d}$ | $27.00 \pm 0.00^{b}$ |
| Earliness (days) | $40.67 \pm 4.81^{bc}$ | $37.25 \pm 1.53^{c}$ | $41.27 \pm 1.16^{bc}$ | $46.00 \pm 2.00^{b}$ | $40.50 \pm 1.44^{bc}$ | $58.67 \pm 5.55^{a}$ |
| Total yield (g) | $215.87 \pm 15.43^{a}$ | $263.84 \pm 46.40^{a}$ | $255.72 \pm 28.04^{a}$ | $134.87 \pm 4.82^{b}$ | $220.75 \pm 3.17^{a}$ | $256.87 \pm 44.34^{a}$ |
| Biological efficiency (%) | $77.26 \pm 5.52^{a}$ | $82.60 \pm 14.53^{a}$ | $56.79 \pm 6.23^{b}$ | $39.73 \pm 1.42^{c}$ | $73.68 \pm 1.06^{a}$ | $71.33 \pm 12.31^{ab}$ |

The use of hardwood sawdust or cereal straw, cotton seed husk, and corn cobs is common in the commercial production of *G. lucidum* [36–38]. However, the highly increased market demand during the last twenty years led to the exploitation of several other substrates, which demonstrated an increase in BE of up to 40% by replacing part of the sawdust by residues such as stillage grain from rice-spirit distilleries or tea wastes [39,40]. An even higher increase (up to 75%) in BE was reported on substrates consisting of maize straw supplemented with wheat and maize bran [41]. Olive by-products were evaluated for the first time as substrates in *G. lucidum* cultivation. Though the BS (control) performed notably better than the other media tested, BS:OLPR 3:1 showed that it is a promising alternative substrate (worth examining in future experiments with additional strains to assess their potential suitability) in *G. lucidum* mushroom production since it supported relatively high yields with very satisfactory earliness values.

On the other hand, *P. ostreatus* colonized most substrates within 26–28 days, but ratios of OLPR exceeding 50% (*w/w*, in respect to WS) delayed substrate colonization by up to 10 days (Table 1). Significant differences were observed in regard to earliness in *P. ostreatus*; mushroom appearance required the most time in WS:OLPR 3:1 (46 days) and WS:TPOMW 1:1 (59 days), whereas primordia formation was noted within 37–41 days in the other treatments examined (Table 1). The evaluation of yield and BE for *P. ostreatus* evidenced that the supplementation of the conventional substrate with OLPR or TPOMW up to 50% had a positive impact on mushroom performance, i.e., an increase in yield ranging from 5 to 49 g $kg^{-1}$ and BE enhancement by up to 5.5% (Table 1). It is noteworthy that only WS:OLPR 1:3 presented a markedly negative effect on *P. ostreatus* mushroom production.

Hence, most of the substrates based on olive by-products supported similar or even better performance in *P. ostreatus* when compared to the control (WS). Both WS:OLPR and WS:TPOMW mixtures performed well, since BE reached 83% and 74%, respectively. This is in agreement with the outcome of previous pertinent studies reporting that addition of TPOMW in rather low ratios (20% *w/w*) resulted in improved *P. ostreatus* mushroom production (by up to 30%), while a further increase could be achieved only after supplementation with composted TPOMW due to the reduced toxicity and higher nutrient availability [21]. In general, the use of different fungal strains in combination with the high variation in the physicochemical properties of olive by-products, which are much affected by soil and climatic conditions, variety, and olive oil extraction process [42,43], results in a considerable variability in the values of mushroom cultivation parameters when such materials are used as substrates, e.g., BE: 50–137% for *P. ostreatus* cultivated on—supplemented or not—olive by-products [10,15,19,21,44].

### 2.3. Assessment of Nutritional Composition of Mushrooms Produced on Olive By-Products

In the frame of the present study, the outcome of proximate analysis on *G. lucidum* mushrooms revealed that the main ingredients (with the exception of crude fat) differed—albeit not always significantly—among fruit-bodies cultivated on various substrates (Table 2). In general, it was observed that mushrooms grown on the control (BS) contained less ash and protein compared to most of those produced on other substrates, while mushrooms from BS:OLPR 1:3 showed the highest content (Table 2). Of additional interest was the significantly higher content in crude fibers detected in fruit-bodies formed on BS:OLPR 3:1 (62 g $kg^{-1}$ d.w.). In addition, no significant differences were noted when crude fat values were compared among treatments. In the case of *P. ostreatus*, the addition of OLPR to WS (i.e., from plain WS to WS:OLPR 1:3) resulted in a progressive reduction in ash, crude fat, and fiber content, as well as in an increase of crude protein content by up to 42% (WS versus WS:OLPR 1:3). Furthermore, small variations were detected in total carbohydrates and gross energy content of both mushroom species.

In general, variations noted in mushroom composition were linked to the use of different substrates, e.g., the increases in ash and protein content of mushrooms were associated with high ratios of TPOMW, which could be attributed to its higher concentration in metals and nitrogen, respectively [21,43]. This outcome is in accordance to previous results obtained in *P. ostreatus* produced on paper scraps [45] or after the supplementation of wood-based substrates of *G lucidum* with tea waste (which is richer than sawdust in nitrogen and minerals [39]). Moreover, a significantly higher protein content was detected in cultivated mushrooms growing on substrates rich in nitrogen, e.g., spent beer grains supplemented with bran, wheat straw mixed with sugar beet, olive leaves mixed with TPOMW for *P. ostreatus* [10,46,47], beech sawdust supplemented with wheat bran for *Hericium americanum* [48], or wheat straw amended with poultry manure or rolled oats and soybean flour for *C. cylindracea* [49,50]. Therefore, nitrogen supplementation, apart of improving mushroom yield, also contributes at obtaining a final product with an elevated protein content.

**Table 2.** Crude composition and bioactive compounds content in *Ganoderma lucidum* and *Pleurotus ostreatus* mushrooms cultivated in six substrates consisting of beech sawdust (BS); wheat straw (WS); two-phase olive mill waste (TPOMW), and olive pruning residues (OLPR) in various mixtures. Values are in g kg$^{-1}$ d.w., except of gross energy (kcal 100 g$^{-1}$ d.w.), total phenolics (mg gallic acid equivalents per g d.w.), antiradical activity (mmol Trolox equivalents per g d.w.), and reducing power (mmol Trolox equivalents per g d.w.), and are expressed as means ± standard deviation of means, $n = 4$. Lack of superscript letters in common indicates statistically significant differences (Gabriel's *t*-test, $p < 0.05$) for comparisons of treatment means between substrates.

| | *Ganoderma lucidum* | | | | | |
|---|---|---|---|---|---|---|
| | **BS** | **BS:OLPR 3:1** | **BS:OLPR 1:1** | **BS:OLPR 1:3** | **BS:TPOMW 3:1** | **BS:TPOMW 1:1** |
| Ash | $3.10 \pm 0.08^{c}$ | $3.08 \pm 0.12^{c}$ | $3.41 \pm 0.33^{c}$ | $5.20 \pm 0.32^{a}$ | $3.68 \pm 0.07^{bc}$ | $4.26 \pm 0.18^{b}$ |
| Crude fiber | $47.93 \pm 1.86^{b}$ | $62.48 \pm 1.18^{a}$ | $43.80 \pm 2.02^{b}$ | $21.57 \pm 9.66^{c}$ | $52.89 \pm 3.77^{ab}$ | $49.34 \pm 1.56^{b}$ |
| Crude fat | $2.21 \pm 0.08^{a}$ | $2.45 \pm 0.05^{a}$ | $2.03 \pm 0.49^{a}$ | $2.04 \pm 0.50^{a}$ | $1.56 \pm 0.52^{a}$ | $1.10 \pm 0.12^{a}$ |
| Crude protein | $16.84 \pm 0.62^{bc}$ | $17.06 \pm 0.15^{bc}$ | $15.28 \pm 0.55^{c}$ | $23.25 \pm 4.08^{abc}$ | $18.85 \pm 0.55^{ab}$ | $22.21 \pm 0.79^{a}$ |
| T. carbohydrates | $77.86 \pm 1.05^{a}$ | $77.41 \pm 2.07^{a}$ | $79.28 \pm 3.47^{a}$ | $69.52 \pm 3.29^{b}$ | $75.92 \pm 1.85^{a}$ | $72.44 \pm 1.38^{ab}$ |
| Gross energy | $399 \pm 13^{a}$ | $400 \pm 17^{a}$ | $397 \pm 4^{a}$ | $389 \pm 11^{a}$ | $393 \pm 2^{a}$ | $388 \pm 3^{a}$ |
| α-glucan | $2.09 \pm 0.48^{b}$ | $1.68 \pm 0.68^{b}$ | $6.84 \pm 1.99^{a}$ | $6.22 \pm 2.27^{a}$ | $3.08 \pm 1.38^{ab}$ | $3.87 \pm 0.30^{ab}$ |
| β-glucan | $35.83 \pm 2.05^{a}$ | $43.10 \pm 6.38^{a}$ | $35.06 \pm 5.14^{a}$ | $31.27 \pm 6.19^{a}$ | $34.72 \pm 1.03^{a}$ | $32.87 \pm 2.00^{a}$ |
| Total phenolics | $3.25 \pm 0.09^{b}$ | $4.23 \pm 0.27^{a}$ | $3.89 \pm 1.15^{ab}$ | $4.02 \pm 0.51^{a}$ | $2.98 \pm 0.31^{b}$ | $2.99 \pm 0.32^{b}$ |
| Antiradical activity | $8.49 \pm 1.32^{a}$ | $9.56 \pm 0.41^{a}$ | $8.46 \pm 1.05^{a}$ | $10.41 \pm 1.37^{a}$ | $7.18 \pm 2.78^{a}$ | $6.37 \pm 0.72^{a}$ |
| Reducing power | $12.99 \pm 1.95^{b}$ | $13.96 \pm 0.41^{b}$ | $18.90 \pm 0.89^{a}$ | $13.76 \pm 4.88^{ab}$ | $12.93 \pm 3.82^{b}$ | $13.26 \pm 3.93^{b}$ |
| | *Pleurotus ostreatus* | | | | | |
| | **WS** | **WS:OLPR 3:1** | **WS:OLPR 1:1** | **WS:OLPR 1:3** | **WS:TPOMW 3:1** | **WS:TPOMW 1:1** |
| Ash | $8.49 \pm 0.82^{a}$ | $7.95 \pm 0.10^{a}$ | $6.49 \pm 0.14^{b}$ | $6.32 \pm 0.45^{b}$ | $8.99 \pm 0.03^{a}$ | $9.42 \pm 1.90^{a}$ |
| Crude fiber | $18.99 \pm 1.97^{a}$ | $17.16 \pm 1.24^{a}$ | $16.54 \pm 1.94^{a}$ | $14.01 \pm 1.34^{ab}$ | $15.47 \pm 1.13^{ab}$ | $12.97 \pm 2.44^{b}$ |
| Crude fat | $2.54 \pm 0.17^{a}$ | $2.47 \pm 0.14^{a}$ | $1.87 \pm 0.04^{b}$ | $1.62 \pm 0.21^{b}$ | $2.74 \pm 0.17^{a}$ | $2.70 \pm 0.21^{a}$ |
| Crude protein | $15.22 \pm 1.29^{c}$ | $16.00 \pm 0.37^{c}$ | $19.88 \pm 2.34^{a}$ | $21.54 \pm 0.24^{a}$ | $17.08 \pm 0.58^{b}$ | $19.32 \pm 0.24^{ab}$ |
| T. carbohydrates | $73.75 \pm 2.81^{a}$ | $73.58 \pm 4.27^{a}$ | $71.76 \pm 0.87^{a}$ | $70.52 \pm 1.46^{a}$ | $71.19 \pm 0.98^{a}$ | $68.56 \pm 3.64^{a}$ |
| Gross energy | $379 \pm 5^{a}$ | $381 \pm 14^{a}$ | $383 \pm 2^{a}$ | $383 \pm 9^{a}$ | $378 \pm 4^{a}$ | $376 \pm 9^{a}$ |
| α-glucan | $8.75 \pm 0.37^{a}$ | $7.25 \pm 1.98^{a}$ | $6.32 \pm 2.07^{a}$ | $5.69 \pm 2.40^{a}$ | $6.00 \pm 1.14^{a}$ | $2.17 \pm 0.58^{b}$ |
| β-glucan | $30.64 \pm 2.45^{a}$ | $28.02 \pm 1.61^{a}$ | $28.87 \pm 3.45^{a}$ | $25.58 \pm 0.21^{a}$ | $31.49 \pm 0.43^{a}$ | $29.56 \pm 2.42^{a}$ |
| Total phenolics | $2.01 \pm 0.11^{c}$ | $3.09 \pm 0.34^{a}$ | $2.73 \pm 0.17^{ab}$ | $2.57 \pm 0.27^{b}$ | $2.81 \pm 0.24^{a}$ | $2.99 \pm 0.34^{a}$ |
| Antiradical activity | $2.49 \pm 0.14^{b}$ | $3.79 \pm 0.47^{ab}$ | $2.81 \pm 0.57^{b}$ | $5.59 \pm 1.21^{a}$ | $3.27 \pm 0.43^{ab}$ | $4.68 \pm 0.67^{a}$ |
| Reducing power | $5.78 \pm 0.88^{a}$ | $6.69 \pm 1.27^{a}$ | $6.85 \pm 1.39^{a}$ | $5.55 \pm 0.22^{a}$ | $3.47 \pm 0.27^{b}$ | $3.94 \pm 0.04^{b}$ |

In regard to glucans, *G. lucidum* generally presented lower content in α-glucan (1.7–6.8%) and higher content in β-glucan (31.3–43.1%) than *P. ostreatus* (2.2–8.8% and 25.6–31.5%, respectively) (Table 2). In terms of α-glucan content, the use of olive by-products exhibited opposite effects on the two species studied. Hence, the addition of OLPR to BS at ratios exceeding 50% resulted in a three-fold increase in α-glucan content in *G. lucidum* (i.e., BS versus BS:OLPR 1:3), whereas the opposite was noted in *P. ostreatus* mushrooms, where increased ratios of the control substrate (WS) favored higher content in α-glucan (8.75% in WS versus. 6.32% and 2.17% in WS:OLPR 1:1 and WS:TPOMW 1:1, respectively). In regard to β-glucans, high content was found in fruit-bodies, while a notable increase (by 20% reaching up to a total of 43%) was recorded in *G. lucidum* mushrooms when cultivated in a BS:OLPR 3:1 substrate (versus the control, BS). However, no significant differences were recorded, which can be attributed to the high variability observed among replicates in BS:OLPR treatments. On the other hand, a similar content in β-glucans (26–31%) was found in *P. ostreatus* fruit-bodies produced in different substrates.

To the best of our knowledge, no data are available concerning qualitative and quantitative variation in mushroom glucans production when different cultivation substrates are used. This is the first time that glucans have been comparatively evaluated in *G. lucidum* fruit-bodies produced on various substrates. Results revealed that the nature of substrate significantly affects relative content in glucans structural types (α- and β- glucan). In addition, *G. lucidum* fruit-bodies were shown to be among the richest in β-glucan content (i.e., 31–43%) when compared to a wide range of mushrooms from various species previously examined [51]. A similar enhancement in β-glucans content in mushrooms cultivated on olive by-products was reported in *H. erinaceus* and *Pleurotus* spp. [18,19] and was attributed to the activation of β-glucan synthase due to the toxicity of olive mill by-products [52]. Studies using soy residues [53,54] or wheat straw [55] also showed that the glucan content in *G. lucidum* is affected by the substrate composition, which, in turn, has an impact on the antimicrobial, antioxidant, and cytotoxic properties of mushroom extracts.

In general, higher concentrations of total phenolics (TPC) and antioxidant activity were recorded in *G. lucidum* mushrooms in respect to *P. ostreatus* (Table 2). Fruit-bodies of both species demonstrated significant increase in TPC when produced on substrates containing OLPR, i.e., by up to 30% and 54% for *G. lucidum* and *P. ostreatus*, respectively (as compared to TPC content in fruit-bodies cultivated in BS and WS, respectively). In contrast, the addition of TPOMW in substrates resulted in TPC increase in *P. ostreatus* mushrooms only, indicating that such interactions also depend on the fungal species used.

Similar trends were observed for the antioxidant activity of fruit-bodies (Table 2). *G. lucidum* mushrooms exhibited increased antiradical activity and reducing power values in BS:OLPR substrates only (significantly different in the case of reducing power, i.e., up to 45% higher than the control). On the other hand, *P. ostreatus* mushrooms showed higher values for antiradical activity only; the respective values in fruit-bodies from olive-based by-products were more than double in respect to the those obtained from the control substrate. Finally, a statistical analysis of the results showed that a significant correlation between TPC and antioxidant activity in *P. ostreatus* only ($r = 0.87$ for TPC versus antiradical activity, and $r = 0.45$ for TPC versus reducing power, $p < 0.05$), demonstrating that phenolics are among the main antioxidant compounds in *Pleurotus* mushrooms. An increase in total phenolics concentration, which was also assessed for *Pleurotus eryngii*, *P. nebrodensis*, and *H. erinaceus* [18,19], seems to be associated with the selective absorption of substrates' organic components—including phenolic and terpenic compounds—by several cultivated mushrooms, and it is closely associated with elevated antioxidant activities [19].

### 2.4. Fourier Transform Infrared (FTIR) Analysis

The analysis of the recorded FTIR spectra revealed a notable/extended variation in *G. lucidum* mushrooms produced in different substrates as opposed to *P. ostreatus* mushrooms, where differences in spectra were mostly observed in the region 1800–400 $cm^{-1}$ (Figure 2). In both species, differences were mainly related to the quantitative changes of compounds present in mushrooms which were consequently associated with the respective region vibrations (Figure 3). The spectral regions at 1670–1610, 1550, and 1240 $cm^{-1}$ are related to the C=O stretching (Amide I band), N–H in-plane/C–N stretching (Amide II band), and C–N stretching (Amide III band) of proteins, respectively, and demonstrate that *G. lucidum* fruit-bodies contains higher amounts of protein than those of *P. ostreatus*, which is in general agreement with measurements of crude protein content (Table 2). In addition, since the region 1180–1000 $cm^{-1}$ could provide information regarding the C–O stretching vibration in polysaccharides and the $PO_2^-$ stretching (1080 $cm^{-1}$) of phospholipids [56], our results demonstrated that *G. lucidum* mushrooms show a higher carbohydrates content (at 1160 $cm^{-1}$) than *P. ostreatus*, which is also in accordance to the crude carbohydrates content measured (Table 2).

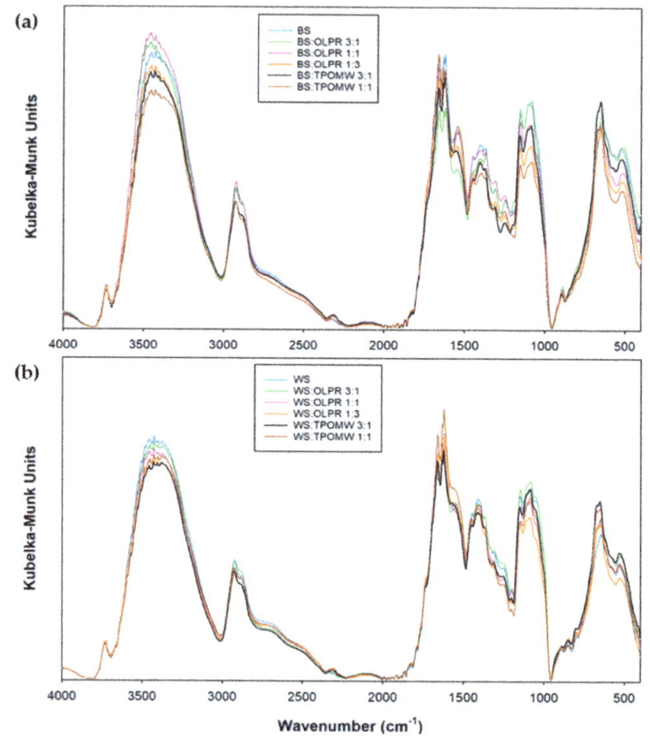


**Figure 2.** FTIR spectra of (**a**) *Ganoderma lucidum* and (**b**) *Pleurotus ostreatus* mushrooms produced on different substrates (and their mixtures), i.e., BS: Beech sawdust, WS: Wheat straw, TPOMW: Two-phase olive mill waste, and OLPR: Olive pruning residues.

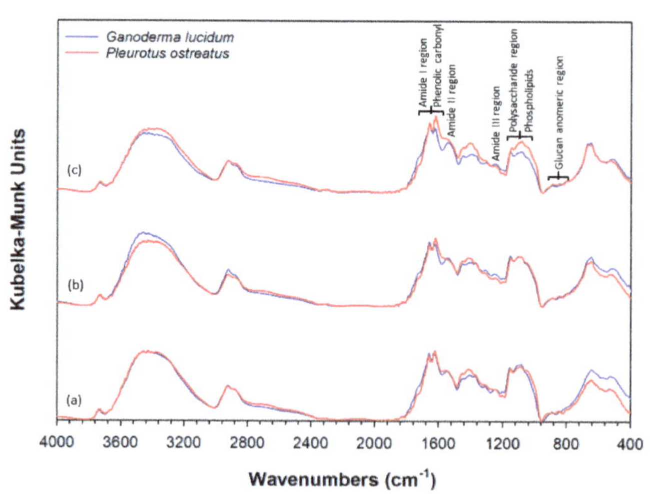


**Figure 3.** Spectroscopic comparison of samples from *Ganoderma lucidum* and *Pleurotus ostreatus* mushrooms produced on (**a**) control substrate (BS and WS, respectively), (**b**) mixture of control substrate and olive pruning (BS:OLPR 1:1 and WS:OLPR 1:1), and (**c**) mixture of control substrate and two-phase olive-mill waste (BS:TPOMW 1:1 and WS:TPOMW 1:1).

A comparison between the different substrates of each species was also performed in order to examine the effect of each substrate on mushroom content in compounds of interest. The region 950–750 $cm^{-1}$ provides information regarding the anomeric region of glucans (Figure 4); more specifically, the region at 890 $cm^{-1}$ for the C–H deformation in β-glucans and the regions at 930 and 850 $cm^{-1}$ for the asymmetric ring vibration and the C–H deformation in α-glucans, respectively [56,57]. In regard to glucans, *G. lucidum* mushroom spectra corresponding to different substrates were in close agreement with the respective content measured by the commercial kit (Table 2). Moreover, due the very low content in α-glucans in most of *G. lucidum* treatments, the peaks related to the α-anomeric structure of glucans were almost absent. However, the ability of FTIR to detect the two substrates (i.e., BS:OLPR 1:1 and BS:OLPR 1:3) that produced mushrooms with relatively higher content in α-glucan was noteworthy. On the other hand, β-glucan in *P. ostreatus* mushrooms was marginally detected, mostly due to the very similar content in β-glucans among treatments and the high—in some cases—standard deviation between replicates in the same treatment. Nevertheless, α-glucan region absorption intensities in *P. ostreatus* followed the previously measured α-glucan content (Table 2).

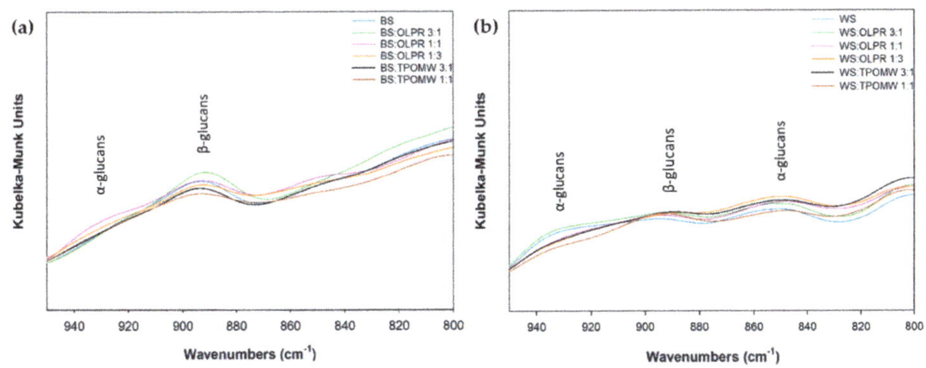


**Figure 4.** Glucans and recorded spectra by diffuse reflectance infrared Fourier transform (DRIFT) spectroscopy in the region 950–800 $cm^{-1}$ of (**a**) *Ganoderma lucidum* and (**b**) *Pleurotus ostreatus* mushrooms produced on different substrates (BS: Beech sawdust, WS: Wheat straw, TPOMW: Two-phase olive mill waste, and OLPR: Olive pruning residues).

Information regarding lipids in cell membranes of mushrooms can be obtained through the FTIR spectrum in three regions and originates from various types of molecular vibrations: (i) The acyl chain vibrations (i.e., $CH_3$ and $CH_2$ asymmetric and symmetric stretching vibrations, as well as $CH_2$ bending and rocking vibrations), (ii) the headgroup vibrations (i.e., $PO_2^-$ stretching vibration), and (iii) the interface regions (i.e., C=O stretching vibration) [56]. However, due to the high content in a wide range of organic compounds, which results in an extended peak overlapping, information regarding the lipid/fat content of the produced mushrooms could be acquired from the region 3000–2750 $cm^{-1}$, and, more specifically, by the peaks at 2930 and 2880 $cm^{-1}$ corresponding to the $CH_2$ asymmetric and $CH_3$ symmetric stretching of acyl chain respectively (Figure 5). For both mushroom species, peak intensity seems to follow the measured crude fat of mushrooms from different substrates (Table 2). However, some differences can be observed, probably due to the very similar fat content of the mushrooms from different treatments and the relatively high standard deviation observed among replicates within individual treatment (Table 2).

Finally, little information could be acquired regarding the phenolic content (and consequently the antioxidant activity) of mushrooms from the recorded spectra. The characteristic regions for phenolic compounds, e.g., 1670–1600 $cm^{-1}$ for the carbonyl vibration [56], are related to proteinic, phenolic, and flavonoid carbonyl, and, therefore, the recorded FTIR intensities in this region cannot be associated with an individual group of compounds.

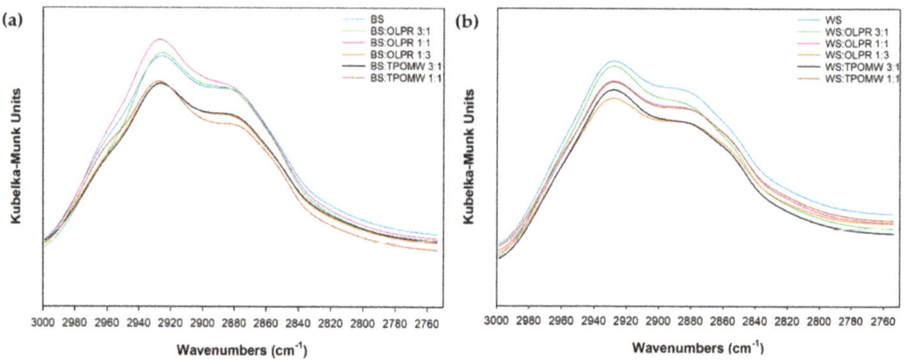


**Figure 5.** Acyl chain vibration of lipids and recorded DRIFT spectra in the region 3000–2750 $cm^{-1}$ of (**a**) *Ganoderma lucidum* and (**b**) *Pleurotus ostreatus* mushrooms produced on different substrates (BS: Beech sawdust, WS: Wheat straw, TPOMW: Two-phase olive mill waste, and OLPR: Olive pruning residues).

### 2.5. Evaluation of Prebiotic Potential of G. lucidum and P. ostreatus Mushrooms

In the present work, the growth of four intestinal bacterial isolates (*Lactobacillus acidophilus*, *L. gasseri*, *Bifidobacterium bifidum*, and *B. longum*) was studied by using lyophilized powder from two edible/medicinal mushrooms (*P. ostreatus* and *G. lucidum*) deriving from different substrates as the sole carbon source. No significant differences were observed in initial inocula (t = 0) of each of the four bacterial strains among all tested substrates (Figure 6a–d). In the case of lactobacilli, mushrooms supported a comparable-to-glucose bacterial growth of both strains after 24 h of incubation (Figure 6a,b), and of *L. gasseri* strain after 48 h of incubation (Figure 6b). Interestingly, in the case of *L. acidophilus*, mushrooms of *P. ostreatus* (from all tested substrates) and *G. lucidum* (from BS:OLRP 1:1) induced a significant increase in bacterial levels compared to glucose after 48 h of incubation, while the rest of *G. lucidum* treatments exhibited similar growth levels compared to standard medium (Figure 6a).

The potential lactogenic effect of polysaccharide extracts from several *Pleurotus* spp. (e.g., *P. ostreatus*, *P. eryngii*, *P. citrinopileatus*, and *P. salmoneo-stramineus*) has been previously reported [30,58]. Furthermore, the growth of lactobacilli strains was supported by extracts of *G. lucidum* based on in vitro fermentation models [26,59,60] and animal studies [31,61]. The previously reported induction of *Lactobacillus* spp. growth by *P. ostreatus* and *G. lucidum* polysaccharides was also evident in the present work based on lyophilized samples of entire fruit-bodies. Our results were also in line with the fact that, in some cases, the growth rates of lactobacilli in mushroom extracts were higher than those of bifidobacteria, which is probably related to the more efficient fermentation profile of *Lactobacillus* spp. [26,58].

A more differentiated, strain-specific pattern of bacterial growth was observed in the case of bifidobacteria after 24 h and 48 h of incubation (Figure 6c,d). In detail, all mushroom treatments used as the sole carbon source-induced growth of *B. longum* within 24 h, similar to that of glucose control; moreover, this effect was also evident only in the case of *G. lucidum* cultured in BS:OLRP 1:1 or BS after a 48 h cultivation of *B. longum* (Figure 6d). In contrast, the growth of *B. bifidum* was significantly hindered in all mushroom substrates after 24 h and 48 h of incubation, with a more drastic effect (i.e., undetectable bacterial levels at both time points) observed in the case of *G. lucidum* in BS:TPOMW 1:1. Furthermore, no detectable growth of *B. bifidum* was recorded in *G. lucidum* BS:OLRP 1:1 and *P. ostreatus* WS:TPOMW 1:1 after 48 h of incubation (Figure 6c).

In the past, probiotic strains of *Bifidobacterium* spp. (e.g., *B. longum*, *B. pseudocatenulatum*) were used in order to determine the prebiotic capacity of extracts from *G. lucidum* [26,59,60] and *Pleurotus* spp. [30,58]. *B. bifidum* was tested for the first time as an indicator of mushroom prebiotic activity in the present study. In vitro and animal-based data have suggested the bifidogenic potential of *G. lucidum* and *Pleurotus* spp. extracts, with an emphasis in the marked variability of the effect exerted by the bacterial strain and by differences in the chemical structure of polysaccharides [58–60]. The strain and substrate specific effects on bifidobacterial growth were also evident in our study, implying that the biological potency of the tested mushrooms could be modified by regulating the formulation of their cultivation substrate to meet the nutritional requirements of the probiotic strains examined. In line with previous data [26], our results suggested a short period of enhanced *B. longum* growth for both mushrooms, with a more prolonged bacterial growth, especially in the case of *G. lucidum* BS:OLRP. This result could be attributed to the presence of simple sugars (e.g., glucose) that are rapidly consumed by bifidobacteria and the substrate-dependent variation in the polysaccharides content of mushrooms [26,62]. In contrast, the *B. bifidum* growth on de Man Rogosa and Sharpe (MRS) culture medium was significantly higher compared to mushroom-based substrates in all cases, indicating that the latter were not suitable for this particular probiotic microorganism.

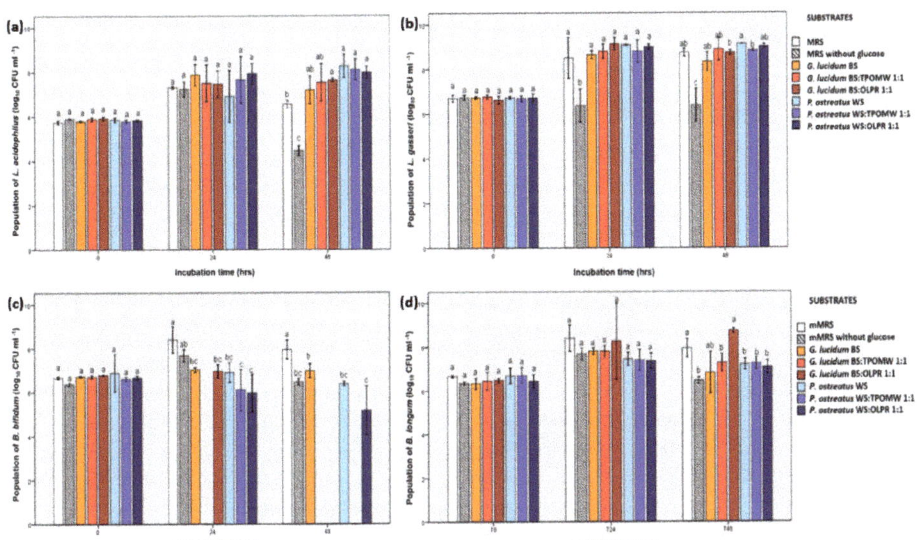


**Figure 6.** Populations of (**a**) *L. acidophilus*, (**b**) *L. gasseri*, (**c**) *B. bifidum*, and (**d**) *B. longum* at inoculation (T0) and after 24 and 48 h of incubation, which were grown on lyophilized *G. lucidum* and *P. ostreatus* mushroom powder as the sole carbon source. Mushrooms were cultivated on conventional substrates (BS and WS for *G. lucidum* and *P. ostreatus*, respectively) and on mixtures with two phase olive mill wastes (TPOMW) and olive tree pruning (OLPR). Columns represent means of bacterial populations ($\log_{10}$ CFU $mL^{-1}$) ± standard errors of means, ($n$ = 4). A lack of letters in common indicates statistically significant differences (Duncan's $t$-test, $p < 0.05$) for comparisons of treatment means between different substrates.

## 3. Materials and Methods

### 3.1. Biological Material

In the frame of this study, *G. lucidum* LGAM 9720 and *P. ostreatus* LGAM 1123 strains were examined. Both strains were previously isolated from the wild (Greece), were routinely maintained on potato dextrose agar (PDA; Conda, Spain), and were preserved in agar slants and as submerged cultures at 4 °C in the fungal Culture Collection of Laboratory of General and Agricultural Microbiology (Agricultural University of Athens, Athens, Greece).

### 3.2. Substrates for Fungal Growth—Determination of Mycelium Growth Rates

The suitability of the following media (mainly composed of olive-based by-products) to serve as substrates for fungal growth was evaluated: (i) Olive pruning residues (OLPR) were used alone or in mixtures with beech sawdust (BS) for *G. lucidum* or with wheat straw (WS) for *P. ostreatus* in ratios of 25, 50, and 75% (*w*/*w*, f.w.); (ii) two phase olive mill waste (TPOMW) and BS or WS were mixed as in case (i); and (iii) TPOMW was mixed with OLPR in ratios of 25 and 50% (*w*/*w*, f.w.); BS and WS alone were used as controls, while the moisture content of the substrates was 52–65%. Olive based by-products were obtained from an olive mill in Kalamata (Peloponnese, southwest Greece), while WS and BS were derived from a wood processing industry in Athens and a mushroom cultivation farm in Evvoia (central Greece), respectively. Glass 'race tubes' (200 × 30 mm) were filled with the aforementioned substrates, sterilized (121 °C, 1.1 atm, 1 h) and then inoculated with a 6 mm diameter agar plug taken from the actively growing periphery of *G. lucidum* or *P. ostreatus* developing on a Petri dish with PDA. Fungal growth took place in an incubation chamber at 25 °C in the dark. Mycelium growth was recorded daily until the substrate was completely colonized, and linear growth rates were calculated as previously described [63].

### 3.3. Mushroom Cultivation Substrates—Assessment of Production Parameters

On the basis of the results obtained through the comparative assessment of mycelium growth rates in 'race tubes' containing the substrates initially tested, five mixtures were selected for further examination in respect to mushroom production (i.e., BS or WS with OLPR in ratios of 3:1, 1:1, and 1:3, and BS or WS in mixtures with TPOMW in ratios of 3:1 and 1:1); BS and WS substrates were used as controls for *G. lucidum* and *P. ostreatus*, respectively.

For the inoculation of mushroom cultivation substrates, cereal grain spawn was prepared according to method previously described [15]. Then, polypropylene autoclavable bags were filled with 1 kg of substrate (moisture content: 50–68%), sterilized (121 °C, 1.1 atm, 1 h), and inoculated with spawn at a 5% *w*/*w* rate. Four replicates per substrate and strain were used. The incubation of substrates was performed at 25 °C in the dark; for primordia formation, the temperature and relative air humidity were set at 16 °C and 95%, respectively, and illumination was provided (700 lux $m^{-2}$, 12h $day^{-1}$ with fluorescent lamps). As soon as primordia were formed, the $CO_2$ level was maintained at 800–1000 ppm, the temperature and relative humidity were set at 18 °C and 85%, respectively, and illumination was increased to 1000 lux $m^{-2}$.

To evaluate the suitability of different substrates for supporting mushroom production, the following parameters were studied: (i) Incubation time, defined as the time between inoculation and complete colonization of the substrate by fungal hyphae; (ii) earliness, defined as the time elapsed between the day of inoculation and the day of primordia appearance, (iii) total yield, expressed as fresh weight of mushrooms harvested; and (iv) biological efficiency, calculated as the percentage ratio of fresh mushroom weight over the dry weight of the substrate. The entire cropping period lasted 90 and 120 days for *P. ostreatus* and *G. lucidum*, respectively.

Following harvest, *P. ostreatus* and *G. lucidum* mushrooms were freeze-dried in a Telstar Cryodos apparatus and milled to fine powder. Pertinent samples were stored at −20 °C until subjected to analyses.

### 3.4. Chemical Analysis of Mushrooms

Ash, crude fiber and crude fat were determined according to methods described by the Association of Official Agricultural Chemists, AOAC International [64]. Nitrogen content was assessed by a CHN elemental analyzer (Carlo Erba, EA1108, Isomass Scientific Inc., Calgary, Canada), and crude protein was calculated by employing the converting factor 4.38 [10]. Nitrogen-free extracts (total carbohydrates) were estimated by the formula: 100 − (moisture + protein + fat + ash contents), and gross energy (kcal

100 $g^{-1}$ f.w.) was calculated according to the equation: Energy = 4 × (g protein + g carbohydrate) + 9 × (g fat) [65].

The Mushroom and Yeast Beta-Glucan assay kit (Megazyme Int., Bray, Ireland) was used for the determination of total and α-glucans content in mushroom samples according to the manufacturer's instructions. The content of β- glucans was calculated by subtracting α-glucans from total glucans.

### 3.5. Measurement of Total Phenolic Content and Antioxidant Activity of Mushrooms Methanolic Extracts

Methanolic extracts were prepared as previously described [66]. In detail, 0.5 g of freeze-dried mushroom samples were extracted with 10 mL of methanol (48 h, 100 rpm at room temperature). Separation was performed by centrifugation in 2500 rpm (10 min), and the precipitate was re-extracted with 2.5 mL of the same solvent for 2 h. Then, the two supernatants were combined, concentrated to 2 mL, and maintained in GS vials tubes in deep freeze. The determination of total phenolic content (TPC) was carried out by the Folin–Ciocalteu method [67]. The antioxidant activity of methanolic extracts were determined as previously described [68] by measuring (i) the radical scavenging activity through the use of the stable free radical molecule DPPH and (ii) the reducing antioxidant potential through the ferric ion reduction activity power (FRAP). The antioxidant activity of mushroom methanolic extracts was evaluated in terms of radical scavenging activity and reducing antioxidant potential using the DPPH and the ferric ion reducing power (FRAP) assays, respectively, as previously described [68]. In the first assay, 0.025 mL of methanolic extract was added in 0.975 mL DPPH (0.1 mM in MeOH) in eppendorf tubes and vortexed. After the mixture was left to stand for 30 min in the dark, the reduction of DPPH was determined by measuring the absorption at 515 nm (U-2001 Spectrophotometer, Hitachi, Tokyo, Japan). DPPH alone served as the blank. In regard to the FRAP assay, 0.05 mL of each extract was added in 0.05 mL $FeCl_3$ solution (3 mM in 5 mM HCl) in an eppendorf tube, and then it was vortexed and incubated at 37 °C for 30 min. Then, 0.9 mL of 2,4,6-tri(2-pyridyl)-1,3,5-triazine (TPTZ) solution (1 mM in 0.05 M HCl) was added, and the absorbance of the product of the reaction between $Fe^{2+}$ and TPTZ was measured at 620 nm against a blank. For the blank, the $FeCl_3$ solution was replaced by distilled water. All assays were performed in triplicate, and the quantifications were based on calibration curves using syringic acid for TPC and Trolox for radical scavenging activity and FRAP.

### 3.6. FTIR Analysis

The FTIR spectra were recorded by a Nicolet 6700 spectrometer (ThermoScientific, Waltham, MA, USA) equipped with a deuterated triglycine sulfate (DTGS) detector (Nichrome source with a potassium bromide beam-splitter) and Omnic 7.3 software. For each freeze-dried sample, 64 scans of the infrared region between 4000 and 400 $cm^{-1}$ at a resolution of 4 $cm^{-1}$ were recorded in triplicates and averaged. Afterwards, the recorded spectra were transformed using the Kubelka–Munk algorithm, which corrects the lack of linearity between the spectral intensity and sample concentration and is applied in diffuse reflectance infrared Fourier transform (DRIFT) spectroscopy where quantitative analysis or relevant comparisons are required. The Kubelka–Munk-transformed spectra were then smoothed by the Savitzky–Golay algorithm [5 points each side (total window of 11 smoothing points) and a zero-order polynomial], adaptively baseline corrected, and normalized by mean. All spectral transformations were performed by the SpectraGryph 1.2.7 software (https://www.effemm2.de/spectragryph/).

### 3.7. Determination of Mushrooms Prebiotic Potential

The mushrooms' prebiotic properties were assessed for each species in three out of the six fungus/substrate combinations examined (and previously presented) by using lyophilized samples from fruit-bodies deriving from: (i) OLPR in mixture (1:1, *w/w*) with BS or WS for *G. lucidum* and *P. ostreatus*, respectively; (ii) TPOMW and BS or WS in the same ratio as in (i); and (iii) BS and WS alone (control substrates) for *G. lucidum* and *P. ostreatus*, respectively.

Four bacterial strains originally isolated from human feces, i.e., two *Lactobacillus* strains (*L. acidophilus* DSM20079 and *L. gasseri* from the Culture Collection of Harokopio University) and two *Bifidobacterium* strains (*B. bifidum* DSM20456 and *B. longum* from the Culture Collection of Harokopio University), were cultivated in Man, de Rogosa, Sharpe (MRS) or modified MRS (mMRS), respectively, under anaerobic conditions at 37 °C. Modified MRS is commonly used for the cultivation of bifidobacteria; it differs from MRS by containing 0.05% HCl-cysteine, a reducing agent which lowers the oxido/reduction potential in culture media to ensure anaerobic conditions [69]. Activated cultures were subsequently inoculated (1% *v/v*) in culture media (MRS, mMRS, with and without glucose: controls) and in culture media with lyophilized mushroom samples as the sole carbon source (2% *w/v*). In order to avoid bacterial contamination from the non-sterile mushrooms, gentamicin was added in the mMRS medium only (4 mg $L^{-1}$). Cultures were incubated under anaerobic conditions at 37 °C for 48 h (Bactron 1.5, SHELLAB, Cornelius, OR, USA). Samples were taken at 0, 24, and 48 h in order to test the viability of the bacterial strains based on plate-count techniques.

### *3.8. Statistical Analysis*

Four replicates for each treatment were used in "race tubes" and in mushroom cultivation experiments. Results are presented as mean ± standard deviation. An analysis of variance followed by a Duncan's *t*-test at 5% level of probability, which was performed for assessing differences between the means of the various substrates examined, while relationships between variables (at significance levels of 0.05 and 0.01) were determined by Pearson's correlation coefficient through the use of SPSS (version 22, IBM, Armonk, NY, USA) software.

## 4. Conclusions

The supplementation *P. ostreatus* substrates by olive by-products increased total mushroom yields and reduced the time required for mushroom formation. In contrast, all alternative substrates had a negative effect on *G. lucidum* cultivation parameters. As concerns the crude composition of mushrooms, high ratios of OLPR to wheat-straw resulted in an increase of crude protein and a reduction of ash, crude fiber, and fat content in *P. ostreatus* fruit-bodies. On the other hand, *G. lucidum* mushrooms exhibited up to a three-fold increase in α-glucan or a significant enhancement of β-glucan content when cultivated on OLPR-based substrates (in comparison to beech sawdust). In addition, several substrates based on olive by-products led to the production of fruit-bodies with increased total phenolic content and antioxidant activity. FTIR spectra confirmed the qualitative/quantitative differentiation of mushrooms composition and demonstrated their suitability as an inexpensive and fast method for determining relevant changes in fruit-bodies content. Moreover, mushroom powder supported/enhanced the growth of *L. acidophilus* and *L. gasseri* after 24 and/or 48 h incubation, while a strain-specific pattern was observed in bifidobacteria; both mushrooms hindered *B. bifidum* growth, and they supported a similar-to-glucose growth for *B. longum*.

**Author Contributions:** Carried out the mushroom cultivation experiments, carried out the analytical and statistical analyses, and wrote the original draft, G.K.; performed the experiments related to the assessment of the prebiotic potential of mushrooms, M.P., E.K.M., and M.K.; wrote the respective part of the manuscript, E.K.M. and A.K.; conducted determination of mushrooms glucan content, carried out the FTIR analysis, and wrote the respective part of the manuscript, G.B.; supervised different aspects of the study, G.I.Z., A.K., and P.A.T.; conceived and designed the experiments, G.I.Z., A.K., V.P., and G.K.; compiled the final version of the manuscript including revision and editing, G.I.Z.; contributed at reviewing and editing, A.K., E.K.M., G.B., P.A.T., V.P., and G.K.; responsible for project administration and funding acquisition, A.K., V.P., and G.I.Z. All authors have read and approved the final submitted manuscript.

**Funding:** This research has been co-financed by the European Union and Greek national funds through the Operational Program Competitiveness, Entrepreneurship and Innovation, under the call RESEARCH—CREATE—INNOVATE (project code: T1EDK-03404).

**Acknowledgments:** Substrates used as controls in the experiments performed were kindly donated by Dirfis Mushrooms IKE (Kathenoi, Evvoia) and Tourikis SA (Athens).

**Conflicts of Interest:** The authors declare no conflict of interest.

## References


1. Kostenidou, E.; Kaltsonoudis, C.; Tsiflikiotou, M.; Louvaris, E.; Russell, L.; Pandis, S. Olive Tree Branches Burning: A major pollution source in the Mediterranean. *Geophys. Res. Abstr.* **2013**, *15*, EGU2013-8298.
2. Roig, A.; Cayuela, M.L.; Sanchez-Monedero, M.A. An overview on olive mill wastes and their valorisation methods. *Waste Manag.* **2006**, *26*, 960–969. [CrossRef] [PubMed]
3. Diamantis, V.; Erguder, T.H.; Aivasidis, A.; Verstraete, W.; Voudrias, E. Wastewater disposal to landfill-sites: A synergistic solution for centralized management of olive mill wastewater and enhanced production of landfill gas. *J. Environ. Manag.* **2013**, *128*, 427–434. [CrossRef]
4. Barbera, A.C.; Maucieri, C.; Cavallaro, V.; Ioppolo, A.; Spagna, G. Effects of spreading olive mill wastewater on soil properties and crops, a review. *Agric. Water Manag.* **2013**, *119*, 43–53. [CrossRef]
5. Chowdhury, A.K.M.M.B.; Akratos, C.S.; Vayenas, D.V.; Pavlou, S. Olive mill waste composting: A review. *Int. Biodeterior. Biodegradation* **2013**, *85*, 108–119. [CrossRef]
6. Ochando-Pulido, J.M.; Pimentel-Moral, S.; Verardo, V.; Martinez-Ferez, A. A focus on advanced physico-chemical processes for olive mill wastewater treatment. *Sep. Purif. Technol.* **2017**, *179*, 161–174. [CrossRef]
7. Pulido, J.M.O. A review on the use of membrane technology and fouling control for olive mill wastewater treatment. *Sci. Total Environ.* **2016**, *563*, 664–675. [CrossRef] [PubMed]
8. Zervakis, G.I.; Koutrotsios, G. Solid state fermentation of plant residues and agro-industrial wastes for the production of medicinal mushrooms. In *Medicinal Plants and Fungi: Recent Advances in Research and Development*; Agrawal, D.C., Tsay, H.-S., Shyur, L.-F., Wu, Y.-C., Wang, S.-Y., Eds.; Springer Nature: Singapore, 2017; pp. 365–396. [CrossRef]
9. Das, N.; Mukherjee, M. Cultivation of *Pleurotus ostreatus* on weed plants. *Bioresour. Technol.* **2007**, *98*, 2723–2726. [CrossRef]
10. Koutrotsios, G.; Mountzouris, K.C.; Chatzipavlidis, I.; Zervakis, G.I. Bioconversion of lignocellulosic residues by *Agrocybe cylindracea* and *Pleurotus ostreatus* mushroom fungi—Assessment of their effect on the final product and spent substrate properties. *Food Chem.* **2014**, *161*, 127–135. [CrossRef]
11. Mandeel, Q.; Al-Laith, A.; Mohamed, S. Cultivation of oyster mushrooms (*Pleurotus* spp.) on various lignocellulosic wastes. *World J. Microbiol. Biotechnol.* **2005**, *21*, 601–607. [CrossRef]
12. Membrillo, I.; Sánchez, C.; Meneses, M.; Favela, E.; Loera, O. Particle geometry affects differentially substrate composition and enzyme profiles by *Pleurotus ostreatus* growing on sugar cane bagasse. *Bioresour. Technol.* **2011**, *102*, 1581–1586. [CrossRef] [PubMed]
13. Obodai, M.; Cleland-Okine, J.; Vowotor, K. Comparative study on the growth and yield of *Pleurotus ostreatus* mushroom on different lignocellulosic by-products. *J. Ind. Microbiol. Biotechnol.* **2003**, *30*, 146–149. [CrossRef] [PubMed]
14. Pant, D.; Reddy, U.G.; Adholeya, A. Cultivation of oyster mushrooms on wheat straw and bagasse substrate amended with distillery effluent. *World J. Microbiol. Biotechnol.* **2006**, *22*, 267–275. [CrossRef]
15. Philippoussis, A.; Zervakis, G.; Diamantopoulou, P. Bioconversion of agricultural lignocellulosic wastes through the cultivation of the edible mushrooms *Agrocybe aegerita, Volvariella volvacea* and *Pleurotus* spp. *World J. Microbiol. Biotechnol.* **2001**, *17*, 191–200. [CrossRef]
16. Salmones, D.; Mata, G.; Waliszewski, K.N. Comparative culturing of *Pleurotus* spp. on coffee pulp and wheat straw: Biomass production and substrate biodegradation. *Bioresour. Technol.* **2005**, *96*, 537–544. [CrossRef] [PubMed]
17. Sánchez, A.; Ysunza, F.; Beltrán-García, M.J.; Esqueda, M. Biodegradation of viticulture wastes by *Pleurotus*: A source of microbial and human food and its potential use in animal feeding. *J. Agric. Food Chem.* **2002**, *50*, 2537–2542. [CrossRef]
18. Koutrotsios, G.; Larou, E.; Mountzouris, K.; Zervakis, G.I. Detoxification of olive mill wastewater and bioconversion of olive crop residues into high-value added biomass by the choice edible mushroom *Hericium erinaceus*. *Appl. Biochem. Biotechnol.* **2016**, *180*, 195–209. [CrossRef]
19. Koutrotsios, G.; Kalogeropoulos, N.; Kaliora, A.C.; Zervakis, G.I. Toward an increased functionality in Oyster (*Pleurotus*) mushrooms produced on grape marc or olive mill wastes serving as sources of bioactive compounds. *J. Agric. Food Chem.* **2018**, *66*, 5971–5983. [CrossRef]

20. Zervakis, G.; Yiatras, P.; Balis, C. Edible mushrooms from olive mill wastes. *Int. Biodeterior. Biodegradation* **1996**, *38*, 237–243. [CrossRef]
21. Zervakis, G.I.; Koutrotsios, G.; Katsaris, P. Composted versus raw olive mill waste as substrates for the production of medicinal mushrooms: An assessment of selected cultivation and quality parameters. *BioMed Res. Int.* **2013**. [CrossRef]
22. Royse, D.J. A global perspective on the high five: *Agaricus, Pleurotus, Lentinula, Auricularia* & *Flammulina*. In Proceedings of the 8th International Conference on Mushroom Biology and Mushroom Products (ICMBMP8), New Delhi, India, 19–22 November 2014; pp. 1–6.
23. Gargano, M.L.; van Griensven, L.J.L.D.; Isikhuemhen, O.S.; Lindequist, U.; Venturella, G.; Wasser, S.P.; Zervakis, G.I. Medicinal mushrooms: Valuable biological resources of high exploitation potential. *Plant Biosyst.* **2017**, *151*, 548–565. [CrossRef]
24. Wasser, S.P. Reishi or ling zhi (*Ganoderma lucidum*). In *Encyclopedia of Dietary Supplements*, 1st ed.; Marcel Dekker, Inc.: New York, NY, USA, 2005; pp. 603–622.
25. Tsapatou, A.; Mitsou, E.K.; Patsou, M.; Koutrotsios, G.; Zervakis, G.I.; Kyriacou, A. Potential prebiotic effect of *Pleurotus ostreatus* and *Ganoderma lucidum* mushrooms on human gut microbiota. In Proceedings of the Gut Microbiota for Health World Summit 2015, Barcelona, Spain, 14–15 March 2015; p. 18.
26. Yamin, S.; Shuhaimi, M.; Arbakariya, A.; Fatimah, A.B.; Khalilah, A.K.; Anas, O.; Yazid, A.M. Effect of *Ganoderma lucidum* polysaccharides on the growth of *Bifidobacterium* spp. as assessed using real-time PCR. *Int. Food Res. J.* **2012**, *19*, 1199–1205.
27. Wasser, S.P. Medicinal mushrooms as a source of antitumor and immunomodulating polysaccharides. *Appl. Microbiol. Biotechnol.* **2002**, *60*, 258–274. [PubMed]
28. Camilli, G.; Tabouret, G.; Quintin, J. The complexity of fungal β-glucan in health and disease: Effects on the mononuclear phagocyte system. *Front. Immunol.* **2018**, *9*, 673. [CrossRef] [PubMed]
29. Chang, C.-J.; Lin, C.-S.; Lu, C.-C.; Martel, J.; Ko, Y.-F.; Ojcius, D.M.; Tseng, S.-F.; Wu, T.-R.; Chen, Y.-Y.M.; Young, J.D.; et al. *Ganoderma lucidum* reduces obesity in mice by modulating the composition of the gut microbiota. *Nat. Commun.* **2015**, *6*, 7489. [CrossRef] [PubMed]
30. Freitas, A.C.; Antunes, M.B.; Rodrigues, D.; Sousa, S.; Amorim, M.; Barroso, M.F.; Carvalho, A.; Ferrador, S.M.; Gomes, A.M. Use of coffee by-products for the cultivation of *Pleurotus citrinopileatus* and *Pleurotus salmoneo-stramineus* and its impact on biological properties of extracts thereof. *Int. J. Food Sci. Technol.* **2018**, *53*, 1914–1924. [CrossRef]
31. Khan, I.; Huang, G.; Li, X.; Leong, W.; Xia, W.; Hsiao, W.L.W. Mushroom polysaccharides from *Ganoderma lucidum* and *Poria cocos* reveal prebiotic functions. *J. Funct. Foods* **2018**, *41*, 191–201. [CrossRef]
32. Wasser, S.P. Current findings, future trends, and unsolved problems in studies of medicinal mushrooms. *Appl. Microbiol. Biotechnol.* **2011**, *89*, 1323–1332. [CrossRef]
33. Zhu, F.; Du, B.; Bian, Z.; Xu, B. Beta-glucans from edible and medicinal mushrooms: Characteristics, physicochemical and biological activities. *J. Food Compost Anal.* **2015**, *41*, 165–173. [CrossRef]
34. Avni, S.; Ezove, N.; Hanani, H.; Yadid, I.; Karpovsky, M.; Hayby, H.; Gover, O.; Hadar, Y.; Schwartz, B.; Danay, O. Olive mill waste enhances α-glucan content in the edible mushroom *Pleurotus eryngii*. *Int. J. Mol. Sci.* **2017**, *18*, 1564. [CrossRef]
35. Koutrotsios, G.; Kalogeropoulos, N.; Stathopoulos, P.; Kaliora, A.; Zervakis, G.I. Bioactive compounds and antioxidant activity exhibit high intraspecific variability in *Pleurotus ostreatus* mushrooms and correlate well with cultivation performance parameters. *World J. Microbiol. Biotechnol.* **2017**, *33*, 98. [CrossRef] [PubMed]
36. Li, S.; Dong, C.; Wen, H.A.; Liu, X. Development of Ling-zhi industry in China–emanated from the artificial cultivation in the Institute of Microbiology, Chinese Academy of Sciences (IMCAS). *Mycology* **2016**, *7*, 74–80. [CrossRef] [PubMed]
37. Xia, Z.; Jiang, J.; He, C.; Liu, M.; Liu, D. Preliminary researches on high-yield cultivation techniques of *Ganoderma*. *Hunan Agric. Sci.* **2003**, *6*, 5658.
38. Zhou, X.-W.; Su, K.-Q.; Zhang, Y.-M. Applied modern biotechnology for cultivation of *Ganoderma* and development of their products. *Appl. Microbiol. Biotechnol.* **2012**, *93*, 941–963. [CrossRef] [PubMed]
39. Peksen, A.; Yakupoglu, G. Tea waste as a supplement for the cultivation of *Ganoderma lucidum*. *World J. Microbiol. Biotechnol.* **2009**, *25*, 611–618. [CrossRef]
40. Yang, F.-C.; Hsieh, C.; Chen, H.-M. Use of stillage grain from a rice-spirit distillery in the solid state fermentation of *Ganoderma lucidum*. *Process Biochem.* **2003**, *39*, 21–26. [CrossRef]

41. Ji, H.; Wang, Q.; Wang, H.; Chen, W.; Zhu, C.; Hou, H.; Zhang, Z. A fundamental research of mushroom cultivation using maize straw. *Edible Fungi China* **2001**, *20*, 10–17.
42. Aggoun, M.; Arhab, R.; Cornu, A.; Portelli, J.; Barkat, M.; Graulet, B. Olive mill wastewater microconstituents composition according to olive variety and extraction process. *Food Chem.* **2016**, *209*, 72–80. [CrossRef] [PubMed]
43. Ntougias, S.; Gaitis, F.; Katsaris, P.; Skoulika, S.; Iliopoulos, N.; Zervakis, G.I. The effects of olives harvest period and production year on olive mill wastewater properties—evaluation of *Pleurotus* strains as bioindicators of the effluent's toxicity. *Chemosphere* **2013**, *92*, 399–405. [CrossRef]
44. Ruiz-Rodriguez, A.; Soler-Rivas, C.; Polonia, I.; Wichers, H.J. Effect of olive mill waste (OMW) supplementation to oyster mushrooms substrates on the cultivation parameters and fruiting bodies quality. *Int. Biodeterior. Biodegradation* **2010**, *64*, 638–645. [CrossRef]
45. Fernandes, Â.; Barros, L.; Martins, A.; Herbert, P.; Ferreira, I.C. Nutritional characterisation of *Pleurotus ostreatus* (Jacq. ex Fr.) P. Kumm. produced using paper scraps as substrate. *Food Chem.* **2015**, *169*, 396–400. [CrossRef] [PubMed]
46. Manzi, P.; Gambelli, L.; Marconi, S.; Vivanti, V.; Pizzoferrato, L. Nutrients in edible mushrooms: An inter-species comparative study. *Food Chem.* **1999**, *65*, 477–482. [CrossRef]
47. Wang, D.; Sakoda, A.; Suzuki, M. Biological efficiency and nutritional value of *Pleurotus ostreatus* cultivated on spent beer grain. *Bioresour. Technol.* **2001**, *78*, 293–300. [CrossRef]
48. Atila, F.; Tüzel, Y.; Faz Cano, A.; Fernandez, J.A. Effect of different lignocellulosic wastes on *Hericium americanum* yield and nutritional characteristics. *J. Sci. Food Agric.* **2017**, *97*, 606–612. [CrossRef] [PubMed]
49. Isikhuemhen, O.S.; Mikiashvili, N.A.; Kelkar, V. Application of solid waste from anaerobic digestion of poultry litter in *Agrocybe aegerita* cultivation: Mushroom production, lignocellulolytic enzymes activity and substrate utilization. *Biodegradation* **2009**, *20*, 351–361. [CrossRef] [PubMed]
50. Uhart, M.; Piscera, J.M.; Albert, E. Utilization of new naturally occurring strains and supplementation to improve the biological efficiency of the edible mushroom *Agrocybe cylindracea*. *J. Ind. Microbiol. Biotechnol.* **2008**, *35*, 595–602. [CrossRef] [PubMed]
51. Sari, M.; Prange, A.; Lelley, J.I.; Hambitzer, R. Screening of beta-glucan contents in commercially cultivated and wild growing mushrooms. *Food Chem.* **2017**, *216*, 45–51. [CrossRef] [PubMed]
52. Malavazi, I.; Goldman, G.H.; Brown, N.A. The importance of connections between the cell wall integrity pathway and the unfolded protein response in filamentous fungi. *Brief. Funct. Genomics* **2014**, *13*, 456–470. [CrossRef]
53. Hsieh, C.; Yang, F.-C. Reusing soy residue for the solid-state fermentation of *Ganoderma lucidum*. *Bioresour. Technol.* **2004**, *91*, 105–109. [CrossRef]
54. Shi, M.; Yang, Y.; Guan, D.; Wang, Y.; Zhang, Z. Evaluation of solid-state fermentation by *Ganoderma lucidum* using soybean curd residue. *Food Bioproc. Tech.* **2013**, *6*, 1856–1867. [CrossRef]
55. Ćilerdžić, J.; Vukojević, J.; Stajić, M.; Stanojković, T.; Glamočlija, J. Biological activity of *Ganoderma lucidum* basidiocarps cultivated on alternative and commercial substrate. *J. Ethnopharmacol.* **2014**, *155*, 312–319. [CrossRef] [PubMed]
56. Socrates, G. *Infrared and Raman Characteristic Group Frequencies: Tables and Charts*, 3rd ed.; John Wily & Sons Ltd.: Chichester, UK, 2001.
57. Synytsya, A.; Novak, M. Structural analysis of glucans. *Ann. Transl. Med.* **2014**, *2*, 17. [PubMed]
58. Synytsya, A.; Míčková, K.; Synytsya, A.; Jablonský, I.; Spěváček, J.; Erban, V.; Kováříková, E.; Čopíková, J. Glucans from fruit bodies of cultivated mushrooms *Pleurotus ostreatus* and *Pleurotus eryngii*: Structure and potential prebiotic activity. *Carbohydr. Polym.* **2009**, *76*, 548–556. [CrossRef]
59. Mohd Hamim, H.M.; Shuhaimi, M.; Yazid, A.M.; Ali, A.M.; Anas, O.M.; Asilah, A.T.; Wahab, M.N.; Shukor, M.Y.A. Growth of probiotic bacteria in trypticase phytone yeast medium supplemented with crude polysaccharides from *Ganoderma lucidum*. *Malays. J. Microbiol.* **2010**, *6*, 47–56. [CrossRef]
60. Liu, Y.H.; Lin, Y.S.; Lin, K.L.; Lu, Y.L.; Chen, C.H.; Chien, M.Y.; Shang, H.F.; Lin, S.Y.; Hou, W.C. Effects of hot-water extracts from *Ganoderma lucidum* residues and solid-state fermentation residues on prebiotic and immune-stimulatory activities in vitro and the powdered residues used as broiler feed additives in vivo. *Bot. Stud.* **2015**, *56*, 17. [CrossRef] [PubMed]

61. Meneses, M.E.; Martínez-Carrera, D.; Torres, N.; Sánchez-Tapia, M.; Aguilar-López, M.; Morales, P.; Sobal, M.; Bernabé, T.; Escudero, H.; Granados-Portillo, O.; et al. Hypocholesterolemic properties and prebiotic effects of Mexican *Ganoderma lucidum* in C57BL/6 mice. *PLoS ONE* **2016**, *11*, e0159631. [CrossRef]
62. Saidou, C.; Tchatchueng, J.B.; Ndjouenkeu, R.; Roux, D. Extraction and partial characterisation of hydrocolloid gums from some African legumes. *Int. J. Food Eng.* **2011**, *7*, 15. [CrossRef]
63. Zervakis, G.; Philippoussis, A.; Ioannidou, S.; Diamantopoulou, P. Mycelium growth kinetics and optimal temperature conditions for the cultivation of edible mushroom species on lignocellulosic substrates. *Folia Microbiol.* **2001**, *46*, 231–234. [CrossRef]
64. AOAC. *Official Methods of Analysis*, 16th ed.; Association of Official Analytical Chemists: Arlington, VA, USA, 1995.
65. Manzi, P.; Marconi, S.; Aguzzi, A.; Pizzoferrato, L. Commercial mushrooms: Nutritional quality and effect of cooking. *Food Chem.* **2004**, *84*, 201–206. [CrossRef]
66. Kalogeropoulos, N.; Yanni, A.E.; Koutrotsios, G.; Aloupi, M. Bioactive microconstituents and antioxidant properties of wild edible mushrooms from the island of Lesvos, Greece. *Food Chem. Toxicol.* **2013**, *55*, 378–385. [CrossRef]
67. Singleton, V.L.; Rossi, J.A. Colorimetry of total phenolics with phosphomolybdic phosphotungstic acid reagents. *Am. J. Enol. Vitic.* **1965**, *16*, 144–158.
68. Arnous, A.; Makris, D.P.; Kefalas, P. Correlation of pigment and flavanol content with antioxidant properties in selected aged regional wines from Greece. *J. Food Compost Anal.* **2002**, *15*, 655–665. [CrossRef]
69. Roy, D. Media for the isolation and enumeration of bifidobacteria in dairy products. *Int. J. Food Microbiol.* **2001**, *69*, 167–182. [CrossRef]

*catalysts*

MDPI

*Article*

# Microalgae Cultivation for the Biotransformation of Birch Wood Hydrolysate and Dairy Effluent

**Sandra Lage [1,†], Nirupa P. Kudahettige [2,†], Lorenza Ferro [3], Leonidas Matsakas [4], Christiane Funk [3], Ulrika Rova [4] and Francesco G. Gentili [1,2,*]**

1 Department of Forest Biomaterials and Technology, Swedish University of Agricultural Sciences, 90183 Umeå, Sweden; sandra.lage@slu.se

2 Department of Wildlife, Fish and Environmental Studies, Swedish University of Agricultural Sciences, 90183 Umeå, Sweden; kudahettide222@yahoo.co.uk

3 Department of Chemistry, Umeå University, 90187 Umeå, Sweden; lorenza.ferro@umu.se (L.F.); christiane.funk@umu.se (C.F.)

4 Biochemical Process Engineering, Division of Chemical Engineering, Department of Civil, Environmental and Natural Resources Engineering, Luleå University of Technology, 97187 Luleå, Sweden; leonidas.matsakas@ltu.se (L.M.); Ulrika.Rova@ltu.se (U.R.)

* Correspondence: francesco.gentili@slu.se; Tel.: +46-090-786-8196

† The authors contributed equally to the work.

Received: 12 December 2018; Accepted: 30 January 2019; Published: 2 February 2019

**Abstract:** In order to investigate environmentally sustainable sources of organic carbon and nutrients, four Nordic green microalgal strains, *Chlorella sorokiniana*, *Chlorella saccharophila*, *Chlorella vulgaris*, and *Coelastrella* sp., were grown on a wood (Silver birch, *Betula pendula*) hydrolysate and dairy effluent mixture. The biomass and lipid production were analysed under mixotrophic, as well as two-stage mixotrophic/heterotrophic regimes. Of all of the species, *Coelastrella* sp. produced the most total lipids per dry weight (~40%) in the mixture of birch hydrolysate and dairy effluent without requiring nutrient (nitrogen, phosphorus, and potassium—NPK) supplementation. Overall, in the absence of NPK, the two-stage mixotrophic/heterotrophic cultivation enhanced the lipid concentration, but reduced the amount of biomass. Culturing microalgae in integrated waste streams under mixotrophic growth regimes is a promising approach for sustainable biofuel production, especially in regions with large seasonal variation in daylight, like northern Sweden. To the best of our knowledge, this is the first report of using a mixture of wood hydrolysate and dairy effluent for the growth and lipid production of microalgae in the literature.

**Keywords:** mixotrophic; heterotrophic; lipids; fatty acid methyl esters; dairy wastewater; birch hydrolysate; green algae; *Coelastrella*; *Chlorella*

## 1. Introduction

Microalgal mass culture has been carried out mainly under photoautotrophic conditions, using light as energy and $CO_2$ as a carbon source [1]. Although metabolite production is relatively high, this cultivation method is frequently associated with low biomass concentrations as a result of the light limitations in the major part of the algal culture [2,3]. Self-shading and/or photoinhibition are common problems in culturing photosynthetic organisms [2,3]. To eliminate the light requirement, microalgae can instead be heterotrophically cultivated to increase cell density and biomass production [4,5]. Additionally, heterotrophic cultivation promotes the accumulation of lipids at the expense of proteins in the biomass, which is a desired feature for biodiesel production from microalgae [6–8]. Species of the genera *Chlorella*, *Tetraselmis*, and *Nitzschia* were shown to grow at higher rates under heterotrophic conditions compared with photoautotrophic systems [9–12]. However, not all microalgae can grow in total darkness. In order to perform heterotrophic growth, microalgal species require special physiological abilities to divide and metabolize in darkness, as they have to rapidly adapt to the new environment and withstand hydrodynamic stresses. For large scale biomass generation, growth in inexpensive and easily sterilized medium is required [12]. A drawback of heterotrophic cultivation is the high production cost of the organic carbon source, a weakness that can be overcome by the use of organic carbon sources recovered from waste streams [13,14].

In mixotrophy, microalgae use light as the main energy source to perform photosynthesis, but both $CO_2$ and organic compounds are equally essential as a carbon source. Depending on the light intensity, concentration of $CO_2$, and availability of organic compounds, the microalgae will either grow photoautotrophically or heterotrophically [15]. Mixotrophy is a suitable culture method for microalgal species that are not able to grow in complete darkness. Although this growth regime is less studied, most microalgal species investigated so far have been shown to produce higher biomass yields along with higher lipid, starch, and protein productivities compared with photoautotrophic regimes [11,16,17]. Therefore, the production of mixotrophic microalgae allows for the integration of photosynthetic and heterotrophic metabolisms during the diurnal cycle, thus reducing the impact of biomass loss during dark-respiration, and decreasing the costs of the organic substances utilised during growth in daylight [18]. For these reasons, mixotrophic cultivation should be preferred within the microalgae-to-biofuels process. Notwithstanding, the cost of the organic carbon source, such as glucose, can account up to 79.3% of the total raw material cost during biodiesel production [14]. To investigate alternative, cheaper carbon and nutrient sources, in this study, we tested a mixture of wood hydrolysate and dairy effluent as a growth medium for the microalgae. The nutrients existing in the dairy effluent can serve as a source of medium nutrient, and the glucose present in the wood hydrolysate can function as an organic carbon source. Dairy effluents and wood hydrolysates are available waste streams in Sweden, and can easily be used as a substrate for the cultivation of microalgae. The dairy industry is generally considered to generate the highest amount of wastewater among the various food processing industries. It is estimated that the production of one litre of milk generates on average between 6 and 10 litres of wastewater [19,20]. The highly diversified processes of this industry, leads to the generation of wastes of diverse quality and quantity. Even though dairy pollutants mainly consist of organic compounds [21], their discharge into freshwater streams can cause pollution problems [20]. Considering that lignocellulose biomasses represents about 50% of the total amount of biomass worldwide [22], it is very relevant from an environmental point of view to recycle different forest residues, such as wood hydrolysate, that consist of sugar-rich fractions comprising derivatives of hemicellulose and cellulose by-products, which can be reused and valorised in a safe and environmental-friendly way [23].

The objectives of this study were as follows: (1) to develop a cheap and effective growth regime (mixotrophic and a two stage mixotrophic/heterotrophic process) for the local green microalgae *Chlorella sorokiniana*, *Chlorella saccharophila*, *Chlorella vulgaris*, and *Coelastrella* sp.; (2) to enhance biomass and lipid production; and (3) to investigate the simultaneous treatment of dairy wastewater and to valorise the hydrolysate from birch wood chips. Although various sources of plant biomass have already been investigated in the literature [24,25], only a few studies have assessed the potential of wood hydrolysate as an organic carbon source for microalgae cultivation [26,27]. To date, the present study is the first to evaluate the feasibility of a mixture of wood hydrolysate and dairy effluent for the growth and lipid production of microalgae.

## 2. Results and Discussion

Four locally isolated microalgal strains (i.e., *C. sorokiniana*, *C. saccharophila*, *C. vulgaris*, and *Coelastrella* sp.) [27] were cultivated in a medium containing birch wood hydrolysate and dairy effluent in the presence or absence of NPK (nitrogen, phosphorus, and potassium) supplementation. The dilution ratio of the birch hydrolysate with dairy effluent was selected in order to provide an adequate nutrient supply to the growth medium (Table 1), and to achieve about a 2 g $L^{-1}$ glucose concentration. Although the microalgae biomass increases when the glucose concentration is increased from 0 to 10 g $L^{-1}$, high glucose concentrations have been shown to negatively affect the lipid accumulation of *Chlorella* species under a mixotrophic condition [28,29]. In addition, when high glucose concentrations are used, most of the glucose is not consumed by the microalgae and remains instead in the medium [29]. For instance, *C. sorokiniana* cultured in a mixotrophic regime for 12 days with 5, 10, and 15 g $L^{-1}$ of glucose produced 0.57 ± 0.06, 0.67 ± 0.07, and 0.53 ± 0.06 g $L^{-1}$ of lipids, respectively, and utilized 93 ± 6, 73 ± 2, and 36 ± 2% (*w*/*w*) of the glucose added [29].

The microalgae production of biomass, lipids, and fatty acid methyl esters (FAMEs) were tested in mixotrophic growth for 7 and 14 days (MT 7d and MT 14d), or during a two-stage growth regime including mixotrophic growth for 7 days, followed by heterotrophic growth (H) for 7 days (MT 7d + H 7d). The two stages' cultivation strategies, including separate steps for growth and for cell stress, have been shown to enhance lipid accumulation in microalgae [30–32]. Therefore, during the first 7 days in mixotrophy, the microalgae were expected to use most of the nutrients and organic carbon, and during the following H 7 days, they were hypothesized to use the remaining organic carbon for lipid accumulation. In MT 14 days, however, the microalgae can photosynthesise during the entire period of time. Both the nutrient and organic carbon limitation in the last 7 days of culture (MT 14d and MT 7d + H 7d) represent a stress factor (Table 1). In medium lacking algae inoculation (control), the total suspended solids were very low (Table 2).

All of the microalgal strains cultivated in the birch hydrolysate and dairy effluent had significantly higher biomass concentrations under MT 14d than under the two-stage mixotrophic/heterotrophic growth, independent of the presence of NPK, with the exception of *C. vulgaris*, where, in the presence of NPK, no statistically significant difference was observed (Figures 1A and 2A), resulting in higher biomass productivities (Table 3). The biomass production during the mixotrophic regime for 7 days (MT 7d), was higher than the mixotrophic regime for 14 days (MT 14d) and of the MT 7d + H 7d, independent of the presence of NPK. The algae already consumed between 50% and 60% of the total organic carbon (TOC) during MT 7d (Table 1), and the TOC consumption was only slightly changed over time. Hence, we speculate that the remaining sugars (i.e., xylose) were not available to algae. Generally, the heterotrophic step had a detrimental effect on the algal biomass, resulting in a lower biomass production, which was particularly evident in *C. sorokiniana* and *C. saccharophila* (Figures 1A and 2A). As observed by others, the dual carbon assimilation (inorganic atmospheric $CO_2$ and organic carbon uptake from the medium) of the mixotrophic microalgae resulted in a higher biomass productivity, which subsequently might lead to the formation of energy storage products as a result of the increased availability of carbon [15,28,30]. Accordingly, the percentage of carbon in the biomass of all of the microalgae strains was slightly higher at MT 14d than at MT 7d + H 7d, which was mainly noticed in the cultures without NPK (Table 3). During MT 7d, a rapid growth of microalgal biomass was facilitated by the presence of sufficient nutrients and carbon (Table 1), while at the stress phase of H 7d, a trophic–metabolic change from a mixotrophic to a heterotrophic process, as well as a potential depletion in organic carbon, resulted in a decrease of biomass production [32]. In the mixotrophic stress phase of MT 14d, in the absence of NPK supplementation, the results were species dependent. NPK supplementation led to higher biomass productivities, independent of the growth regime (Table 3). In the presence of NPK, *C. sorokiniana* produced the highest biomass (Figure 2A, 1.84 ± 0.02 (MT 7d), 1.97 ± 0.03 (MT 14d), and 1.38 ± 0.02 g $L^{-1}$ (MT 7d + H 7d)).

**Table 1.** Nutrients, ammonium ($NH_4^+$-N), nitrate ($NO_3^-$-N), total phosphorous (TP), and total organic carbon (TOC), concentrations (mg $L^{-1}$) in the birch hydrolysate and dairy effluent medium, with and without nitrogen, phosphorus, and potassium (NPK) supplementation, at day 0, day 7 (mixotrophic growth—MT 7d), and day 14 (mixotrophic and two stage mixotrophic/heterotrophic growth (H)—MT 14d and MT 7d + H 7d) in the cultures of *C. sorokiniana*, *C. saccharophila*, *C. vulgaris*, and *Coelastrella* sp. Values are expressed as mean ± standard deviation ($n$ = 2).

| | | Without NPK Supplementation | | | | | With NPK Supplementation | | | | |
|---|---|---|---|---|---|---|---|---|---|---|---|
| | | $NH_4^+$-N | $NO_3^-$-N | TP | TOC (mg/L) | TOC Reduction (%) | $NH_4^+$-N | $NO_3^-$-N | TP | TOC (mg/L) | TOC Reduction (%) |
| | Day 0 | 72.70 ± 0.00 | ≤0.50 | 8.56 ± 0.01 | 663 ± 89 | - | 98.40 ± 1.60 | 247.60 ± 2.00 | 16.15 ± 0.15 | 654 ± 66 | - |
| *C. sorokiniana* | MT 7d | 0.85 ± 0.00 | 7.07 ± 0.07 | 1.58 ± 0.04 | 282 ± 20 | 57.5 | 6.58 ± 0.03 | 5.48 ± 0.12 | 4.15 ± 0.00 | 250 ± 27 | 61.8 |
| | MT 14d | 2.12 ± 0.03 | 5.85 ± 0.19 | 4.49 ± 0.07 | 232.5 ± 23.5 | 64.9 | 1.67 ± 0.01 | 5.00 ± 0.01 | 6.15 ± 0.04 | 198 ± 3 | 69.7 |
| | MT 7d + H 7d | 1.47 ± 0.00 | 6.10 ± 0.53 | 2.20 ± 0.09 | 245 ± 30 | 63 | 39.15 ± 0.15 | 4.91 ± 0.16 | 4.00 ± 0.01 | 202 ± 8 | 69.1 |
| *C. saccharophila* | MT 7d | 0.33 ± 0.00 | 6.64 ± 0.15 | 1.52 ± 0.03 | 312 ± 38 | 52.9 | 13.65 ± 0.05 | 5.50 ± 0.13 | 3.67 ± 0.02 | 234.5 ± 14.5 | 64.1 |
| | MT 14d | 1.52 ± 0.05 | 5.82 ± 0.24 | 3.81 ± 0.09 | 258 ± 22 | 61.1 | 1.28 ± 0.02 | 4.86 ± 0.13 | 6.97 ± 0.30 | 269 ± 11 | 58.9 |
| | MT 7d + H 7d | 0.84 ± 0.00 | 5.68 ± 0.52 | 1.87 ± 0.20 | 241.5 ± 2.5 | 63.6 | 54.70 ± 0.40 | 4.76 ± 0.06 | 5.63 ± 0.01 | 229 ± 18.5 | 65.0 |
| *C. vulgaris* | MT 7d | 6.69 ± 0.02 | 7.00 ± 0.20 | 1.65 ± 0.06 | 298 ± 32 | 55.1 | 8.88 ± 0.03 | 6.89 ± 0.20 | 3.05 ± 0.00 | 221.5 ± 7.5 | 66.1 |
| | MT 14d | 0.96 ± 0.00 | 5.39 ± 0.22 | 4.70 ± 0.12 | 239.5 ± 0.5 | 63.9 | 20.05 ± 0.15 | 6.30 ± 0.15 | 8.03 ± 0.67 | 200.5 ± 9.5 | 69.3 |
| | MT 7d + H 7d | 0.42 ± 0.00 | 6.10 ± 0.57 | 1.64 ± 0.22 | 268.5 ± 9.5 | 59.5 | 15.35 ± 0.05 | 5.08 ± 0.02 | 3.43 ± 0.04 | 187.5 ± 13.5 | 71.3 |
| *Coelastrella* sp. | MT 7d | 0.41 ± 0.00 | 6.88 ± 0.14 | 1.37 ± 0.04 | 304 ± 23.5 | 54.1 | 9.30 ± 0.04 | 42.95 ± 0.55 | 2.98 ± 0.01 | 258.5 ± 2.5 | 60.5 |
| | MT 14d | 0.42 ± 0.00 | 5.86 ± 0.22 | 3.44 ± 0.08 | 274 ± 22 | 58.7 | 3.92 ± 0.04 | 16.15 ± 1.95 | 6.75 ± 0.12 | 227.5 ± 2.5 | 65.2 |
| | MT 7d + H 7d | 1.97 ± 0.01 | 5.47 ± 0.44 | 2.15 ± 0.21 | 231.5 ± 17.5 | 65.1 | 0.78 ± 0.00 | 5.12 ± 0.03 | 5.31 ± 0.01 | 197 ± 9 | 69.9 |

**Table 2.** Total suspended solids (TSS) concentration (mg $L^{-1}$) of controls, that is, a birch hydrolysate and dairy effluent medium without microalgae inoculation, with and without NPK supplementation, cultivated for 7 and 14 days mixotrophically (MT 7d and MT 14d, respectively), or for 7 days mixotrophically and then another 7 days heterotrophically (MT 7d + H 7d). Values are expressed as mean ± standard deviation ($n$ = 4).

| Supplementation | Growth Regime | TSS (mg $L^{-1}$) |
|---|---|---|
| With NPK | MT 7d | 15 ± 0.0 |
| | MT 14d | 27 ± 1.0 |
| | MT 7d + H 7d | 16 ± 0.0 |
| Without NPK | MT 7d | 16 ± 1.0 |
| | MT 14d | 15 ± 0.0 |
| | MT 7d + H 7d | 27 ± 1.0 |

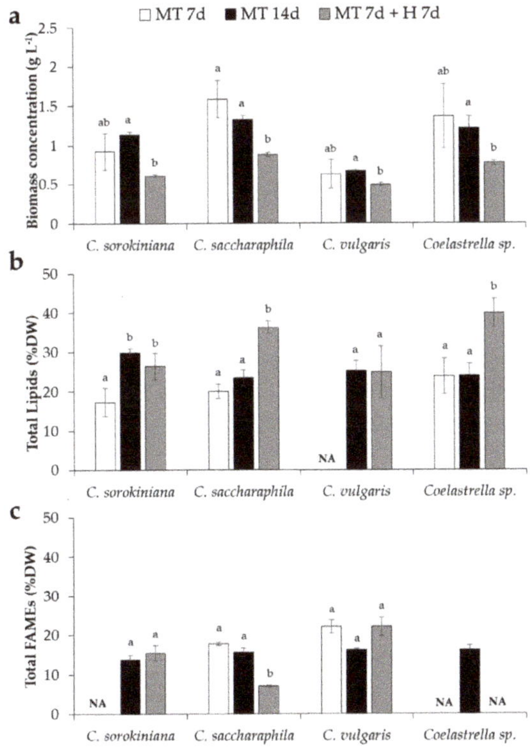


**Figure 1.** (**a**) Biomass concentration (g $L^{-1}$) ($n = 4$); (**b**) total lipids (% dry weight–DW), $n = 4$; and (**c**) total fatty acid methyl esters (FAMEs) (%DW) ($n = 2$), of *C. sorokiniana*, *C. saccharophila*, *C. vulgaris*, and *Coelastrella* sp., grown in a birch hydrolysate and dairy effluent mixture without nitrogen, phosphorus, and potassium (NPK) supplementation. The cells were cultured mixotrophic (MT) for 7 and 14 days (MT 7d and MT 14d), or mixotrophic for 7 days, and then heterotrophic (H) for another 7 days (MT 7d + H 7d). Error bars express the standard deviation of the mean. The different letters above the bars of the same microalgae indicate a significant difference ($p < 0.05$). NA—not available.

In contrast to biomass, the total lipid contents of all of the four strains were generally slightly higher in the absence of NPK than in its presence, independent of the growth regime (Figures 1B and 2B). Nutrient limitation without NPK (Table 1), particularly nitrogen, is known to result in the cessation of microalgal growth, and subsequently, a low biomass content [33,34], but it stimulates the accumulation of reserve lipids, mainly in the form of triacylglicerols (TAGs) [8,30,35]. During photoautotrophic and mixotrophic photosynthesis, the algae assimilate $CO_2$, which is used for growth (proteins) or is stored as carbohydrates, under nutrient limiting conditions. However, TAGs accumulated [30] as nitrogen are not available for protein synthesis [36]. These data are corroborated by the lower total nitrogen content measured in our microalgae grown in the absence of NPK (Table 3), with the exception of *C. vulgaris* at MT 7d + H 7d; and by the nitrogen concentrations in the medium at the end of the experiments (Table 1). Our results further support the connection between microalgal lipid-production and nitrogen availability. However, the microalgal response to nitrogen deficiency is highly variable and strain-specific. Microalgal strains can respond to nutrient starvation by either a several folds increase of lipids, or no change at all, or they can even slightly reduce their lipid amount. Some strains of *Chlorella* were found to accumulate starch during nitrogen starvation, whereas others accumulated mainly neutral lipids [37]. Although, in the absence of NPK, the two-stage MT 7d+ H 7d and MT 14d cultivations resulted in an equal or higher lipid concentration than the MT 7d (not significant for all strains and growth regimes), as reported in the literature [30–32]; the lipid productivities of the MT 7d+ H 7d and MT 14d cultivations were lower than MT 7d. The reduction of lipid productivity for MT 7d+ H 7d and MT 14d was more noticeable in the presence of NPK. The lipid productivities of MT 7d+ H 7d and MT 14d were equivalent, suggesting that the low lipid productivities were a consequence of the low biomass produced during the last 7 days of culture. In the absence of NPK, the prolonged mixotrophic growth for 14 days instead of 7 days only led to a significant increase of lipid content in *C. sorokiniana*; its FAMEs content increased as well, but it was not statistically significant (Figure 1B,C). Again, as observed for the lipids, an increase in FAMEs content did not result in higher FAMEs productivities (Table 3). *C. saccharophila* and *Coelastrella* sp., however, accumulated significantly more lipids in the two-stage mixotrophic/heterotrophic cultivation system (36.32% ± 1.67% and 39.92% ± 3.68% DW, respectively) than in restricted mixotrophic regimes (MT 7d: 20.01% ± 1.87% and 23.77% ± 4.49%; MT 14d: 23.77% ± 4.49% and 23.93% ± 2.99% DW) (Figure 1B). The switch from mixotrophic to heterotrophic metabolism, which induces FAMEs production, is associated with the degradation of structural membranes, photosynthetic proteins, and chlorophylls in the chloroplast [38,39]. In *C. saccharophila* grown in the absence of NPK, the FAMEs' concentration at MT 7d + H 7d was significantly lower than both of the mixotrophic regimes.

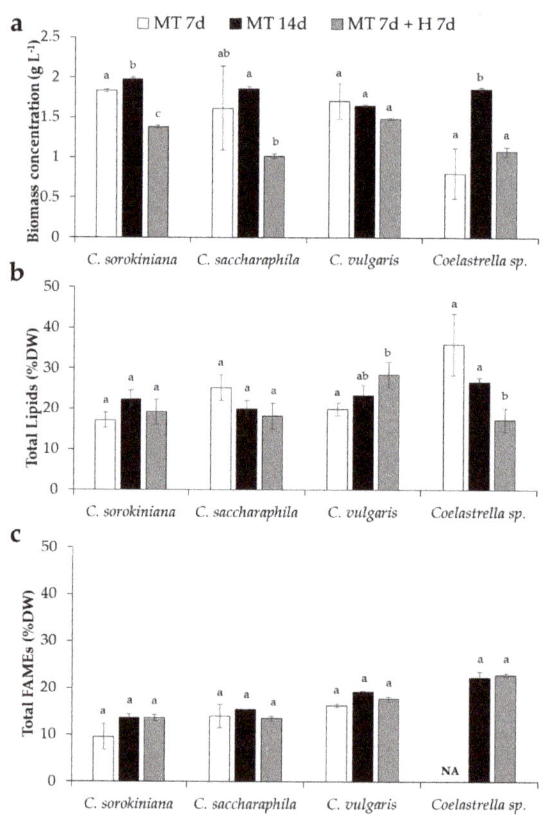


**Figure 2.** (**a**) Biomass concentration (g $L^{-1}$) ($n$ = 4); (**b**) total lipids (%DW) ($n$ = 4); and (**c**) total FAMEs (%DW) ($n$ = 2), of *C. sorokiniana*, *C. saccharophila*, *C. vulgaris*, and *Coelastrella* sp. grown in a birch hydrolysate and dairy effluent with NPK supplementation. The cells were cultured as mixotrophic for 7 and 14 days (MT 7d and MT 14d), or mixotrophic for 7 days, and then heterotrophic for another 7 days (MT 7d + H 7d). Error bars express the standard deviation of the mean. Different letters above the bars of the same microalgae indicate a significant difference ($p < 0.05$). NA—not available.

**Table 3.** Biomass (g $L^{-1}$ $d^{-1}$), lipids (mg $L^{-1}$ $d^{-1}$), and fatty acid methyl esters' (FAMEs) (mg $L^{-1}$ $d^{-1}$) productivities of *C. sorokiniana*, *C. saccharophila*, *C. vulgaris*, and *Coelastrella* sp. cultured in a birch hydrolysate and dairy effluent medium, with and without NPK supplementation (mixotrophic growth, MT 7d, and 14 days of mixotrophic and two stage mixotrophic/heterotrophic growth, MT 14d and MT 7d + H 7d). Values are expressed as mean ± standard deviation ($n$ = 4 (biomass and lipids) and $n$ = 2 (FAMEs)). NA—not available.

| | | Without NPK Supplementation | | | With NPK Supplementation | | |
|---|---|---|---|---|---|---|---|
| | | Biomass | Lipids | FAMEs | Biomass | Lipids | FAMEs |
| *C. sorokiniana* | MT 7d | 0.13 ± 0.03 | 24.75 ± 4.16 | NA | 0.26 ± 0.00 | 24.50 ± 2.33 | 13.71 ± 3.94 |
| | MT 14d | 0.06 ± 0.04 | 21.36 ± 0.66 | 9.79 ± 0.82 | 0.14 ± 0.00 | 15.97 ± 1.40 | 9.75 ± 0.56 |
| | MT 7d + H 7d | 0.04 ± 0.00 | 18.85 ± 2.07 | 11.03 ± 1.36 | 0.10 ± 0.00 | 13.77 ± 1.94 | 9.74 ± 0.48 |
| *C. saccharophila* | MT 7d | 0.23 ± 0.29 | 25.58 ± 2.31 | 25.37 ± 0.56 | 0.23 ± 0.06 | 31.51 ± 8.66 | 19.96 ± 3.50 |
| | MT 14d | 0.09 ± 0.05 | 16.73 ± 1.25 | 11.09 ± 0.73 | 0.13 ± 0.00 | 14.21 ± 1.35 | 11.05 ± 0.00 |
| | MT 7d + H 7d | 0.06 ± 0.00 | 25.94 ± 1.03 | 5.00 ± 0.15 | 0.07 ± 0.00 | 12.98 ± 1.93 | 9.61 ± 0.38 |
| *C. vulgaris* | MT 7d | 0.09 ± 0.02 | NA | 31.57 ± 2.39 | 0.24 ± 0.03 | 28.36 ± 1.97 | 23.20 ± 0.58 |
| | MT 14d | 0.05 ± 0.00 | 17.98 ± 1.60 | 11.54 ± 0.25 | 0.12 ± 0.00 | 16.63 ± 1.61 | 13.70 ± 0.10 |
| | MT 7d + H 7d | 0.04 ± 0.00 | 17.76 ± 4.08 | 15.83 ± 1.69 | 0.11 ± 0.00 | 20.33 ± 1.95 | 12.60 ± 0.34 |
| *Coelastrella* sp. | MT 7d | 0.19 ± 0.05 | 33.95 ± 5.56 | NA | 0.11 ± 0.04 | 41.80 ± 8.13 | NA |
| | MT 14d | 0.09 ± 0.01 | 17.10 ± 1.85 | 11.63 ± 0.79 | 0.13 ± 0.00 | 19.02 ± 0.70 | 15.89 ± 0.90 |
| | MT 7d + H 7d | 0.06 ± 0.00 | 28.51 ± 2.27 | NA | 0.08 ± 0.00 | 12.30 ± 1.82 | 16.28 ± 0.22 |

Notably, in the presence of NPK, the lipid contents of *Coelastrella* sp. were significantly lower during the two-stage mixotrophic/heterotrophic growth compared with the mixotrophic growth (Figure 2B). Thus, this suggests that the excess of carbon absorbed by dual carbon assimilation during mixotrophy resulted in a higher lipid production in the *Coelastrella* sp. mixotrophic cultures [28]. Accordingly, the percentage of carbon in *Coelastrella* sp. was 3% higher in the MT 14d than in the biomass after the MT 7d + H 7d growth. The total lipid content of the *C. vulgaris* biomass was significantly higher after the two-stage growth than in MT 7d in the presence of NPK. However, the *C. vulgaris* FAMEs' concentrations remained unchanged under this condition (Figure 2B,C).

At the end of MT 7d, all of the microalgae strains cultured without NPK were nitrogen-limited, that is, $NH_4^+$-N was reduced 91 to 99% after 7 days of mixotrophic growth (Table 1). $NH_4^+$-N is the preferred nitrogen source for microalgae, mainly because it is the most energetically efficient source, as less energy is required for its uptake. When $NH_4^+$-N and $NO_3^-$-N are supplied together, *Chlorella* sp. uses $NH_4^+$-N first, which is incorporated into the organic compounds produced [9]. In the absence of NPK, the total phosphorous (TP) used for the all microalgae strains growth and development was between 81% and 84% at MT 7d, 45% and 60% at MT 14d, and 74% and 81% at MT 7d + H 7d. In the presence of NPK, the nutrients were also consumed to a great extent in the first 7 days of mixotrophic growth in all of the strains, the $NH_4^+$-N concentration was reduced by 86% to 93%, and $NO_3^-$-N by 98%, and the TP content depleted by between 74% and 82% (Table 1). Nitrogen accounts for 1%–10% of the dry matter in microalgae, and is the most important nutrient affecting growth and lipid accumulation [40]. The nitrogen content in the biomass of the investigated Nordic species was the lowest in the *Coelastrella* sp. grown mixotrophically in the absence of NPK (MT 14d: 3.38% ± 0.11% DW) and highest in *C. vulgaris* grown without NPK in the mixo-/hetero-trophic two-stage regime (MT 7d + H 7d: 11.04% ± 0.03% DW) (Table 4). It is interesting to notice that independent of NPK addition, the algae grown under MT 7d + H 7d had a higher nitrogen concentration and a lower C/N ratio than the algae grown at a mixotrophic condition (MT 14; Table 4). The algal nitrogen content therefore seems to be linked not only to the amount of available nitrogen in the medium (Table 1), but also to the growth regime. Concerning the total carbon content, only slight variations were observed in the different growth regimes in the presence NPK. The highest carbon content was measured in the mixotrophic algal cultures (MT 14d) in the absence of NPK (Table 4). Phosphorous, another essential component for microalgal growth and development, only accounts for about 1% of the total microalgal biomass (approximately 0.3%–0.6%) [41]. The results obtained in the present study are in agreement with previous studies on microalgae grown on a different lignocellulosic biomass [24,27]. *C. sorokiniana* grown in 12% beech (*Fagus sylvatica*) wood acid hydrolysate, with approximately 0.37 g $L^{-1}$ organic carbon (glucose + acetate), produced 0.33 ± 0.01 and 0.22 ± 0.01 g $L^{-1}$ biomass in 32.5 h, under mixotrophic and heterotrophic regimes, respectively [27]. The total fatty acid content was 5.20% ± 0.18% and 4.44% ± 0.24% DW under mixotrophic and heterotrophic regimes, respectively [27]. *C. protothecoides* has been found to produce 2.83 g $L^{-1}$ of biomass with a 56.3% DW lipid content after 60 h of mixotrophic cultivation on a plant biomass (rice straw, *Oryza sativa*) hydrolysate (10 g $L^{-1}$ glucose concentration) [24]. In comparison, this microalgae, grown heterotrophically in another plant (cassava, *Manihot esculenta*), a hydrolysate (10 g $L^{-1}$ glucose concentration), for 240 h, produced approximately 7 g $L^{-1}$ of biomass, but only a 22% DW lipid content [25]. In another study of heterotrophic cultivation, *Auxenochlorella protothecoides* was grown for 120 h using Silver birch (*Betula pendula*) and Norway spruce (*Picea abies*) hydrolysates, with a 10 times higher glucose concentration than in the present study. The biomass concentration and lipid content were 8.56 ± 0.21 and 8.37 ± 0.13 g $L^{-1}$, and 66.00% ± 0.33% and 63.08% ± 0.71% DW for the birch and spruce hydrolysate, respectively [26].

**Table 4.** Percentage of total nitrogen (N) and total carbon (C) per DW, as well as the C/N ratio of *C. sorokiniana*, *C. saccharophila*, *C. vulgaris*, and *Coelastrella* sp. grown in a birch hydrolysate and dairy effluent with or without NPK supplementation. The species were cultivated for 14 days mixotrophically (MT 14d), or for 7 days mixotrophically, and then another 7 days heterotrophically (MT 7d + H 7d). Values are expressed as mean ± standard deviation (*n* = 3).

| Species | Supplementation | Growth Regime | N% | C% | C/N Ratio |
|---|---|---|---|---|---|
| *C. sorokiniana* | With NPK | MT 14d | 6.90 ± 0.34 | 50.73 ± 0.45 | 7.37 ± 0.41 |
| | | MT 7d + H 7d | 9.04 ± 0.57 | 49.81 ± 2.06 | 5.52 ± 0.18 |
| | Without NPK | MT 14d | 3.89 ± 0.88 | 56.03 ± 1.37 | 15.34 ± 4.23 |
| | | MT 7d + H 7d | 8.13 ± 0.10 | 51.10 ± 1.35 | 6.28 ± 0.09 |
| *C. saccharophila* | With NPK | MT 14d | 6.84 ± 0.14 | 51.50 ± 0.46 | 7.53 ± 0.18 |
| | | MT 7d + H 7d | 9.33 ± 0.06 | 50.27 ± 0.06 | 5.38 ± 0.03 |
| | Without NPK | MT 14d | 3.77 ± 0.17 | 56.34 ± 0.06 | 14.98 ± 0.69 |
| | | MT 7d + H 7d | 5.54 ± 1.01 | 55.99 ± 0.33 | 10.42 ± 1.70 |
| *C. vulgaris* | With NPK | MT 14d | 7.15 ± 0.07 | 52.86 ± 0.21 | 7.39 ± 0.10 |
| | | MT 7d + H 7d | 8.70 ± 0.19 | 52.59 ± 1.02 | 6.05 ± 0.01 |
| | Without NPK | MT 14d | 7.21 ± 0.34 | 53.82 ± 0.14 | 7.49 ± 0.33 |
| | | MT 7d + H 7d | 11.04 ± 0.03 | 52.47 ± 0.15 | 4.73 ± 0.02 |
| *Coelastrella* sp. | With NPK | MT 14d | 6.75 ± 0.14 | 50.66 ± 0.91 | 7.51 ± 0.12 |
| | | MT 7d + H 7d | 7.59 ± 0.34 | 52.47 ± 0.72 | 6.93 ± 0.40 |
| | Without NPK | MT 14d | 3.38 ± 0.11 | 58.18 ± 0.62 | 17.21 ± 0.17 |
| | | MT 7d + H 7d | 5.87 ± 0.22 | 55.27 ± 1.57 | 9.43 ± 0.64 |

Similar to the FAMEs yields, the FAMEs compositions did also not vary significantly in dependence to the growth regime (Figure 3); the dominant FAMEs were C16:0, C18:1, and C18:2 in all four of the strains, with the amount of C18:3 lower than expected. Generally, the FAMEs' composition of the three *Chlorella* strains was similar, with some minor differences in the C18 FAMEs proportions (Figure 3). C16:0 was the dominant FAME in the three *Chlorella* strains (on average 32% of total FAMEs); in *Coelastrella* sp., methyl oleate (C18:1) was the dominant fatty acid (on average 43%). Interestingly, substantial differences in the FAME composition were observed upon the availability of NPK. In the presence of NPK, the amount of C18:2 increased in all of the strains and growth regimes (Figure 3). Decreased nutrient availability in the absence of NPK (Table 1), and two-stage mixo-/hetero-trophic growth, resulted in a higher relative content of C16:0 and C18:1, while the relative content of C18:3, and potentially other n-3 polyunsaturated fatty acids [42], was lower (Figure 3).

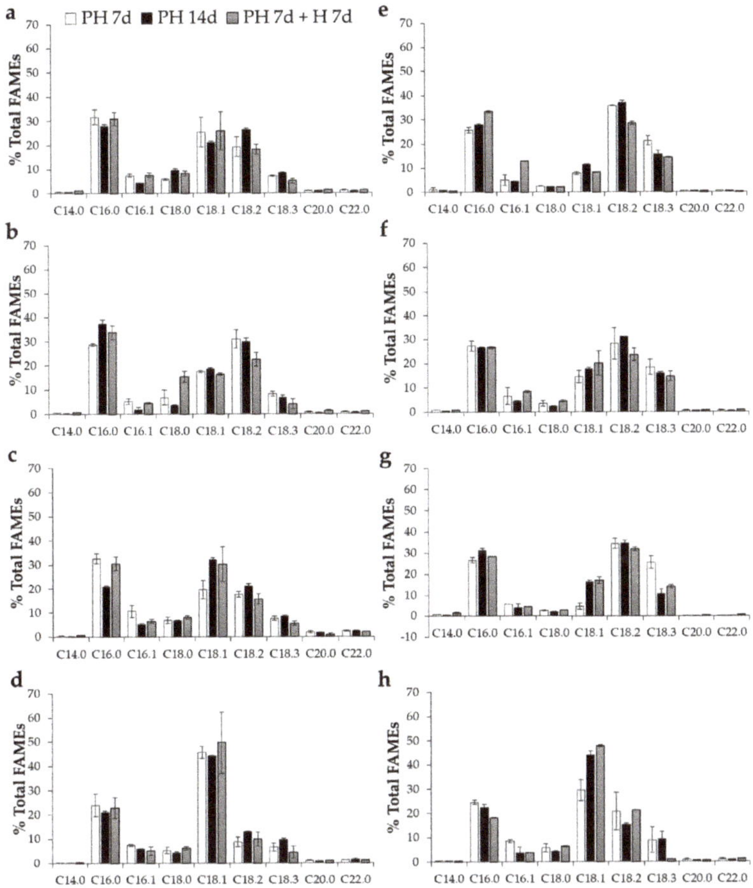


**Figure 3.** FAMEs composition (% total FAMEs) of *C. sorokiniana*, *C. saccharophila*, *C. vulgaris*, and *Coelastrella* sp. grown in birch hydrolysate and dairy effluent, either in the absence (**a**–**d**) or presence (**e**–**h**) of NPK. Cells were grown mixotrophic for 7 and 14 days (MT 7d and MT 14d), or first mixotrophic for 7 days, and then heterotrophic for another 7 days (MT 7d + H 7d). Error bars express the standard deviation of the mean ($n$ = 2).

## 3. Materials and Methods

### 3.1. Collection and Cultivation of Microalgal Strains

Four microalgal strains were isolated from the municipal wastewater (Vakin AB) located in Umeå (63°86′ N), northern Sweden, and were genetically identified as *Chlorella sorokiniana*, *Chlorella saccharophila*, *Chlorella vulgaris*, and *Coelastrella* sp. [27]. The microalgal strains were grown for 7 days in a BG11 medium with 1.5% agar [43] under a 16:8 h light–dark cycle at 22 °C (light) and 16 °C (dark) in a growth cabinet (Conviron A1000 IN, Winnipeg, MB, Canada). Light intensity, expressed as PAR (photosynthetic active radiation), was ≈ 150 μmol $m^{-2}$ $s^{-1}$. Liquid cultures were prepared by inoculating two loops (2-mm diameter loop) of microalgae in Erlenmeyer flasks of 250 mL total volume, containing 150 mL of mixed birch hydrolysate and dairy effluent (see below for preparation), under the same culture conditions described above, and kept under continuous magnetic stirring at 100 rpm. The Erlenmeyer flasks were covered with aluminium foil. Glassware and the substrate were sterilized by autoclaving them at 120 °C for 20 min.

### 3.2. Assessment of Birch Hydrolysate and Dairy Effluent as Growth Medium

Wood chips from Silver birch (*Betula pendula*) were pre-treated with acid catalysed assisted hydrothermal pre-treatment [43]. Briefly, the pre-treatment took place at 190 °C at a holding time of 4–6 min, and sulfur dioxide at a concentration of 0.025 kg $kg_{biomass}^{-1}$, was used as catalyst. As a result, a slurry of approximately 21.69% (*w*/*w*) solid content [43] was produced. The pH of the slurry was adjusted to 5 prior to enzymatic saccharification, and the slurry was diluted to a solid concentration of 20% (*w*/*w*) with a concentrated $Na_2HPO_4$–citric acid buffer, to achieve a final buffer concentration of 50 mM in the diluted slurry. For the enzymatic saccharification, the enzyme solution Cellic® CTec2 (with an enzyme activity of 238 FPU/ml (filter paper unit; [44])) from Novozymes A/S (Bagsværd, Denmark), was used at an enzyme load corresponding to 15 FPU $g^{-1}$ of solids. Saccharification took place in an orbital shaker at 50 °C and 160 rpm for 24 h. At the end of the saccharification, the sugar concentration in the slurry was determined by high-performance liquid chromatography (HPLC; PerkinElmer, Waltham, MA, USA) equipped with a refractive index detector, and a Bio-Rad Aminex HPX-87P column (BioRad, Hercules, CA, USA) operating at 85 °C with 0.6 mL/min of ultrapure water. The sugars' concentration was to be 61.7 g $L^{-1}$ glucose and 42.4 g $L^{-1}$ xylose. The undiluted hydrolysates contained 14.7 g $L^{-1}$ acetic acid, 1.7 g $L^{-1}$ furfural, 0.2 g $L^{-1}$ HMF (determined by HPLC equipped with RI and a Bio-Rad Aminex HPX-87H column operating at 65 °C with 0.6 mL/min of 5 mM $H_2SO_4$), and 4.7 g $L^{-1}$ phenols (determined with the Folin–Ciocalteu method with gallic acid as standard, as described before [45]). The dairy effluent had a glucose concentration of 0.05 g $L^{-1}$, which was also determined by HPLC.

After saccharification, the birch hydrolysate possessed a very dark brown colour, which limited light penetration, and subsequently microalgae photosynthetic activity (data not shown). Additionally, a previous study determined that wood hydrolysate loadings up to 48% inhibited *C. sorokiniana* growth, potentially because of the toxicity of wood hydrolysates towards microalgae [27]. Therefore, the birch hydrolysate was diluted with deionized water, at a dilution ratio of 1:6 (*v*/*v*). After dilution, the pH was adjusted to 7. The diluted birch hydrolysate was mixed with dairy effluent (Norrmejerier, Umeå, Sweden) so as to provide an adequate nutrient supply to the growth medium (Table 1), and to achieve about a 2 g $L^{-1}$ glucose concentration. The birch hydrolysate and dairy effluent mixture was subsequently filtered at room temperature with two layers of paper towel (100% cellulose), with a water filtration velocity of about 1.3 mL $cm^{-2}$ $min^{-1}$ to remove the largest particles, followed by overnight sedimentation at 4 °C. The feasibility of a birch hydrolysate and dairy effluent mixture as a growth medium for microalgae was investigated in the presence or absence of NPK, which was added as sodium nitrate ($NaNO_3$) and dipotassium phosphate ($K_2HPO_4$) to a final concentration of 1.5 and 0.04 g $L^{-1}$, respectively (Table 1).

The four microalgal strains, *Coelastrella* sp., *C. sorokiniana*, *C. saccharophila*, and *C. vulgaris*, were cultivated under either mixotrophic culture conditions, samples were taken after 7 and 14 days (designated as MT 7d and MT 14d, respectively), or for 7 days under mixotrophic, followed by 7 days under heterotrophic culture conditions (designated as MT 7d and H 7d). The cultures were kept with continuous magnetic stirring at 100 rpm and a 16:8 h light–dark cycle at 22 °C (light) and 16 °C (dark) in a growth cabinet (Conviron A1000 IN, Winnipeg, MB, Canada), with PAR of $\approx$ 150 µmol $m^{-2}$ $s^{-1}$ for the mixotrophic conditions, or total darkness for the heterotrophic conditions. Samples of 10 mL were harvested by centrifugation at 3520 g for 5 min, and the obtained pellets were used to analyse the biomass, total lipids, and FAMEs. The supernatants were stored at −20 °C for nutrient analysis. The control samples deficient of microalgae were taken at MT 7d, MT 14d, and MT 7d + H 7d time points from both mediums of birch hydrolysate/dairy effluent, in the presence or absence of NPK, and were processed similar to the samples. The biomass and total lipid analyses were performed in four replicates, and FAMEs and nutrients analyses in duplicates. The total nitrogen (N) and carbon (C) were determined in triplicates.

### 3.3. Analytical Methods

#### 3.3.1. Nutrients Analyses

The nutrients—ammonium ($NH_4^+$-N), nitrate ($NO_3^-$-N), total phosphorus (TP), and total organic carbon (TOC)—were analysed using the commercially available nutrient analyses kits, according to the manufactures' instructions (Hach Lange, Germany). The absorbance measurements were performed using a DR3900 spectrophotometer (Hach Lange, Germany). All of the supernatant samples were thawed at room temperature before the analyses and were diluted, when necessary, to achieve the concentrations within the kit range.

#### 3.3.2. Biomass Concentration

The microalgal biomass concentration was determined in all of the culture conditions immediately after harvesting. The dry weight of the microalgal pellets was determined gravimetrically after oven drying (Memmert, Schwabach, Germany) at 65 °C overnight. The microalgal biomass concentration was expressed as dry weight g $L^{-1}$.

#### 3.3.3. Total Lipids Extraction

The total lipids were extracted from a fresh microalgal biomass using the method of Folch et al. [46], simplified as described in Axelsson and Gentili [47]. The lipids were extracted using a mixture of chloroform, methanol, and NaCl (0.73% in water) (2:1:0.8 $v/v/v$). The recovered lipid phase was vacuum dried in a multi-evaporator (Syncore®Polyvap, Büchi Labortechnik AB, Flawil, Switzerland) at 40 °C, 120 rpm, and 275 mbar for 3 h. The quantity of total lipids was measured gravimetrically, and was expressed as a dry weight percentage.

#### 3.3.4. Fatty Acid Methyl Esters Analysis

The fresh microalgal cells were pelletized and boiled immediately after harvesting in 2 mL of isopropanol at 80 °C for 10 min under gaseous nitrogen atmosphere, and were stored at −20 °C until further analyses. After the total lipid extraction, the fatty acids were isolated and purified on thin layer chromatography (TLC), and subsequently transmethylated into FAMEs according to Lage and Gentili [48], based on Christie and Han [49]. The FAMEs extracts were re-suspended with heptane and injected into a TRACE™ 1310 (Thermo Fisher Scientific, Hägersten, Sweden) GC system equipped with flame ionization detector, and a 30 m FAMEWAX column (Restek Corporation, Bellefonte, Pennsylvania, USA) [48]. The FAMEs were identified by comparison of the retention times with authentic standards. The FAMEs' concentrations were calculated as weight percent by applying theoretical correction factors, and being normalized against the internal standard pentadecanoic acid methyl ester (C15:0).

#### 3.3.5. Total N and Total C Analyses

The total nitrogen and total carbon measurements were performed at the Department of Forest Ecology and Management, Swedish University of Agricultural Sciences (Umeå, Sweden), as described by Werner et al. [50]. The samples were analysed by elemental analyser-isotope ratio mass spectrometry (EA-IRMS). The instrumental setup consisted of an elemental analyser (Flash EA 2000) connected to a continuous flow isotope ratio mass spectrometer (DeltaV), both of which were from Thermo Fisher Scientific (Bremen, Germany). Each sequence of samples was analysed together with two in-house standards in several replicates. The accepted standard deviation of the in-house laboratory standards was <0.15%. The data were corrected for drift and size before the final results were given.

### 3.4. Statistical Analysis

In order to investigate the statistical differences between the biomass, total lipid, and FAMEs' concentrations means of different medium and growth regimes, analysis of variance (one-way ANOVA) followed by post-hoc Student's t-test with Bonferroni correction was applied. Analyses were performed with Microsoft Office Excel 2013 Analysis ToolPak.

## 4. Conclusions

This study shows, for the first time, that birch hydrolysate and dairy effluent can be used, in combination, as an organic carbon and nutrient source for cultivating microalgae in order to produce lipids. All of the microalgae tested, namely *C. sorokiniana*, *C. saccharophila*, *C. vulgaris*, and *Coelastrella* sp., could grow in this medium mixture, under both mixotrophic and two-stage mixotrophic/heterotrophic regimes, independently of NPK supplementation. In comparison to the growth phase alone (MT 7d), the prolonged (14 days instead of 7 days) two-stage cultivation strategy (growth phase followed by stress phase), that is, MT 7d + H 7d and MT 14d, generally resulted in a reduction of the biomass, lipids, and FAMEs' productivity. An exception was *C. saccharophila*, which had a slightly higher lipid productivity at MT 7d + H 7d. For instance, *Coelastrella* sp., the highest lipid producer of this study, accumulated up to 40% DW of the total lipids in the absence of NPK. In the presence of NPK, the nutrient replete condition of the medium impaired the effect of the stress phase. In conclusion, culturing microalgae in waste streams under mixotrophy has the potential to become a successful strategy for microalgal cultivation in northern Sweden; a region with large seasonal variation in daylight availability.

**Author Contributions:** S.L. carried out the analytical and statistical analysis and wrote the original draft. N.P.K. performed the experiment and wrote part of the manuscript. L.F. isolated and genetically identified the local microalgal strains. L.M. prepared the birch hydrolysate and performed glucose HPLC analysis. C.F., U.R., and F.G.G. supervised different aspects of the study. F.G.G. conceived and designed the experiments. All of the authors contributed to the manuscript reviewing and editing. All of the authors have read and approved the final manuscript.

**Funding:** This research was funded by the Kempe foundation (JCK-1512), the Swedish Energy Agency (project number. 38239-1), the Formas–Swedish Research Council for Sustainable Development (Project number 942-2015-92), the Interreg programme Botnia-Atlantica (TransAlgae project), and Bio4Energy—a strategic research environment appointed by the Swedish government.

**Acknowledgments:** The authors are thankful to the technical staff at Norrmejerier for providing the dairy wastewater.

**Conflicts of Interest:** The authors declare no conflict of interest. The funders had no role in the design of the study; in the collection, analyses, or interpretation of data; in the writing of the manuscript; or in the decision to publish the results.

## References

1. Gouveia, L.; Oliveira, A.C. Microalgae as a raw material for biofuels production. *J. Ind. Microbiol. Biotechnol.* **2009**, *36*, 269–274. [CrossRef] [PubMed]
2. Chen, G.Q.; Chen, F. Growing phototrophic cells without light. *Biotechnol. Lett.* **2006**, *28*, 607–616. [CrossRef] [PubMed]
3. Tredici, M.R. Photobiology of microalgae mass cultures: Understanding the tools for the next green revolution. *Biofuels* **2010**, *1*, 143–162. [CrossRef]
4. Huang, G.; Chen, F.; Wei, D.; Zhang, X.; Chen, G. Biodiesel production by microalgal biotechnology. *Appl. Energy* **2010**, *87*, 38–46. [CrossRef]
5. Tabernero, A.; del Valle, E.M.M.; Galán, M.A. Evaluating the industrial potential of biodiesel from a microalgae heterotrophic culture: Scale-up and economics. *Biochem. Eng. J.* **2012**, *63*, 104–115. [CrossRef]
6. Molina Grima, E.; Belarbi, E.H.; Acién Fernández, F.G.; Robles Medina, A.; Chisti, Y. Recovery of microalgal biomass and metabolites: Process options and economics. *Biotechnol. Adv.* **2003**, *20*, 491–515. [CrossRef]

7. Olaizola, M. Commercial development of microalgal biotechnology: From the test tube to the marketplace. *Biomol. Eng.* **2003**, *20*, 459–466. [CrossRef]
8. Miao, X.; Wu, Q. Biodiesel production from heterotrophic microalgal oil. *Bioresour. Technol.* **2006**, *97*, 841–846. [CrossRef]
9. Shi, X.M.; Jiang, Y.; Chen, F. High-yield production of lutein by the green microalga *Chlorella protothecoides* in heterotrophic fed-batch culture. *Biotechnol. Prog.* **2002**, *18*, 723–727. [CrossRef]
10. Qtae, J.; Eun, J.C.; Doo, W.P.; Benoît, V. Sterol dynamics of heterotrophic *Tetraselmis suecica* and its nutritional implication in the bivalve aquaculture. *Aquac. Res.* **2004**, *35*, 371–377. [CrossRef]
11. Wen, Z.-Y.; Chen, F. Production potential of eicosapentaenoic acid by the diatom *Nitzschia laevis*. *Biotechnol. Lett.* **2000**, *22*, 727–733. [CrossRef]
12. Wen, Z.Y.; Chen, F. Heterotrophic production of eicosapentaenoic acid by microalgae. *Biotechnol. Adv.* **2003**, *21*, 273–294. [CrossRef]
13. Perez-Garcia, O.; Escalante, F.M.; de-Bashan, L.E.; Bashan, Y. Heterotrophic cultures of microalgae: Metabolism and potential products. *Water Res.* **2011**, *45*, 11–36. [CrossRef] [PubMed]
14. Koutinas, A.A.; Chatzifragkou, A.; Kopsahelis, N.; Papanikolaou, S.; Kookos, I.K. Design and techno-economic evaluation of microbial oil production as a renewable resource for biodiesel and oleochemical production. *Fuel* **2014**, *116*, 566–577. [CrossRef]
15. Chojnacka, K.; Marquez-Rocha, F.J. Kinetic and stoichiometric relationships of the energy and carbon metabolism in the culture of microalgae. *Biotechnology* **2004**, *3*, 21–34.
16. Heredia-Arroyo, T.; Wei, W.; Ruan, R.; Hu, B. Mixotrophic cultivation of *Chlorella vulgaris* and its potential application for the oil accumulation from non-sugar materials. *Biomass Bioenergy* **2011**, *35*, 2245–2253. [CrossRef]
17. Heredia-Arroyo, T.; Wei, W.; Hu, B. Oil accumulation via heterotrophic/mixotrophic *Chlorella protothecoides*. *Appl. Biochem. Biotechnol.* **2010**, *162*, 1978–1995. [CrossRef]
18. Bassi, A.; Saxena, P.; Aguirre, A.-M. Mixotrophic algae cultivation for energy production and other applications. In *Algal Biorefineries*; Springer: Dordrecht, The Netherlands, 2014; pp. 177–202.
19. Kaur, A.; Vats, S.; Rekhi, S.; Bhardwaj, A.; Goel, J.; Goel, J.; Tanwar, R.S.; Gaur, K.K. Physico-chemical analysis of the industrial effluents and their impact on the soil microflora. *Procedia Environ. Sci.* **2010**, *2*, 595–599. [CrossRef]
20. Russell, P. Effluent and waste water treatment. *Milk Ind. Int.* **1998**, *100*, 36–39.
21. Kushwaha, J.P.; Srivastava, V.C.; Mall, I.D. An overview of various technologies for the treatment of dairy wastewaters. *Crit. Rev. Food Sci. Nutr.* **2011**, *51*, 442–452. [CrossRef]
22. Mancini, G.; Papirio, S.; Lens, P.N.L.; Esposito, G. Solvent Pretreatments of Lignocellulosic Materials to Enhance Biogas Production: A Review. *Energy Fuels* **2016**, *30*, 1892–1903. [CrossRef]
23. Saadatmand, S.; Edlund, U.; Albertsson, A.-C.; Danielsson, S.; Dahlman, O.; Karlstrom, K. Turning hardwood dissolving pulp polysaccharide residual material into barrier packaging. *Biomacromolecules* **2013**, *14*, 2929–2936. [CrossRef]
24. Li, P.; Miao, X.; Li, R.; Zhong, J. In situ biodiesel production from fast-growing and high oil content *Chlorella pyrenoidosa* in rice straw hydrolysate. *BioMed Res. Int.* **2011**, *2011*, 141207.
25. Wei, A.; Zhang, X.; Wei, D.; Chen, G.; Wu, Q.; Yang, S.-T. Effects of cassava starch hydrolysate on cell growth and lipid accumulation of the heterotrophic microalgae *Chlorella protothecoides*. *J. Ind. Microbiol. Biotechnol.* **2009**, *36*, 1383. [CrossRef]
26. Patel, A.; Matsakas, L.; Rova, U.; Christakopoulos, P. Heterotrophic cultivation of *Auxenochlorella protothecoides* using forest biomass as a feedstock for sustainable biodiesel production. *Biotechnol. Biofuels* **2018**, *11*, 169. [CrossRef]
27. Miazek, K.; Remacle, C.; Richel, A.; Goffin, D. Beech wood *Fagus sylvatica* dilute-acid hydrolysate as a feedstock to support *Chlorella sorokiniana* biomass, fatty acid and pigment production. *Bioresour. Technol.* **2017**, *230*, 122–131. [CrossRef]
28. Cheirsilp, B.; Torpee, S. Enhanced growth and lipid production of microalgae under mixotrophic culture condition: Effect of light intensity, glucose concentration and fed-batch cultivation. *Bioresour. Technol.* **2012**, *110*, 510–516. [CrossRef] [PubMed]
29. Wan, M.; Liu, P.; Xia, J.; Rosenberg, J.N.; Oyler, G.A.; Betenbaugh, M.J.; Nie, Z.; Qiu, G. The effect of mixotrophy on microalgal growth, lipid content, and expression levels of three pathway genes in *Chlorella sorokiniana*. *Appl. Microbiol. Biotechnol.* **2011**, *91*, 835–844. [CrossRef]

30. Devi, M.P.; Subhash, G.V.; Mohan, S.V. Heterotrophic cultivation of mixed microalgae for lipid accumulation and wastewater treatment during sequential growth and starvation phases: Effect of nutrient supplementation. *Renew. Energy* **2012**, *43*, 276–283. [CrossRef]
31. Farooq, W.; Lee, Y.C.; Ryu, B.G.; Kim, B.H.; Kim, H.S.; Choi, Y.E.; Yang, J.W. Two-stage cultivation of two *Chlorella* sp. strains by simultaneous treatment of brewery wastewater and maximizing lipid productivity. *Bioresour. Technol.* **2013**, *132*, 230–238. [CrossRef] [PubMed]
32. Chiranjeevi, P.; Venkata Mohan, S. Diverse acidogenic effluents as feedstock for microalgae cultivation: Dual phase metabolic transition on biomass growth and lipid synthesis. *Bioresour. Technol.* **2017**, *242*, 191–196. [CrossRef] [PubMed]
33. Mata, T.M.; Martins, A.A.; Caetano, N.S. Microalgae for biodiesel production and other applications: A review. *Renew. Sustain. Energy Rev.* **2010**, *14*, 217–232. [CrossRef]
34. Minhas, A.K.; Hodgson, P.; Barrow, C.J.; Sashidhar, B.; Adholeya, A. The isolation and identification of new microalgal strains producing oil and carotenoid simultaneously with biofuel potential. *Bioresour. Technol.* **2016**, *211*, 556–565. [CrossRef] [PubMed]
35. Juneja, A.; Ceballos, R.M.; Murthy, G.S. Effects of environmental factors and nutrient availability on the biochemical composition of algae for biofuels production: A review. *Energies* **2013**, *6*, 4607–4638. [CrossRef]
36. Fábregas, J.; Maseda, A.; Domínguez, A.; Ferreira, M.; Otero, A. Changes in the cell composition of the marine microalga, *Nannochloropsis gaditana*, during a light: Dark cycle. *Biotechnol. Lett.* **2002**, *24*, 1699–1703. [CrossRef]
37. Hu, Q.; Richmond, A. *Handbook of Microalgal Culture: Biotechnology and Applied Phycology*; Blackwell Science: Oxford, UK, 2004.
38. Aoki, S.; Matsuka, M.; Hase, E. De-and re-generation of chloroplasts in the cells of *Chlorella protothecoides*: V. Degeneration of chloroplasts induced by different carbon sources, and effects of some antimetabolites upon the process induced by glucose. *Plant Cell Physiol.* **1965**, *6*, 487–498.
39. Hörtensteiner, S.; Chinner, J.; Matile, P.; Thomas, H.; Donnison, I.S. Chlorophyll breakdown in *Chlorella protothecoides*: Characterization of degreening and cloning of degreening-related genes. *Plant Mol. Biol.* **2000**, *42*, 439–450. [CrossRef]
40. Wijffels, R.H.; Barbosa, M.J.; Eppink, M.H. Microalgae for the production of bulk chemicals and biofuels. *Biofuelsbioproducts Biorefining* **2010**, *4*, 287–295. [CrossRef]
41. Hannon, M.; Gimpel, J.; Tran, M.; Rasala, B.; Mayfield, S. Biofuels from algae: Challenges and potential. *Biofuels* **2010**, *1*, 763–784. [CrossRef]
42. Reitan, K.I.; Rainuzzo, J.R.; Olsen, Y. Effect of nutrient limitation on fatty acid and lipid content of marine microalgae. *J. Phycol.* **1994**, *30*, 972–979. [CrossRef]
43. Matsakas, L.; Rova, U.; Christakopoulos, P. Sequential parametric optimization of methane production from different sources of forest raw material. *Front. Microbiol.* **2015**, 6. [CrossRef] [PubMed]
44. Matsakas, L.; Nitsos, C.; Vörös, D.; Rova, U.; Christakopoulos, P. High-Titer Methane from Organosolv-Pretreated Spruce and Birch. *Energies* **2017**, *10*, 263. [CrossRef]
45. Matsakas, L.; Novak, K.; Enman, J.; Christakopoulos, P.; Rova, U. Acetate-detoxification of wood hydrolysates with alkali tolerant *Bacillus* sp. as a strategy to enhance the lipid production from *Rhodosporidium toruloides*. *Bioresour. Technol.* **2017**, *242*, 287–294. [CrossRef] [PubMed]
46. Folch, J.; Lees, M.; Sloane Stanley, G. A simple method for the isolation and purification of total lipids from animal tissues. *J. Biol. Chem.* **1957**, *226*, 497–509. [PubMed]
47. Axelsson, M.; Gentili, F. A single-step method for rapid extraction of total lipids from green microalgae. *PLoS ONE* **2014**, *9*, e89643. [CrossRef] [PubMed]
48. Lage, S.; Gentili, F.G. Quantification and characterisation of fatty acid methyl esters in microalgae: Comparison of pretreatment and purification methods. *Bioresour. Technol.* **2018**, *257*, 121–128. [CrossRef] [PubMed]
49. Christie, W.; Han, X. *Lipid Analysis-Isolation, Separation, Identification and Lipidomic Analysis*; Oily Press: Bridgwater, UK, 2010; 446p.
50. Werner, R.A.; Bruch, B.A.; Brand, W.A. ConFlo III—An interface for high precision δ13C and δ15N analysis with an extended dynamic range. *Rapid Commun. Mass Spectrom.* **1999**, *13*, 1237–1241. [CrossRef]

*catalysts*

MDPI

*Article*

# High Cell Density Conversion of Hydrolysed Waste Cooking Oil Fatty Acids Into Medium Chain Length Polyhydroxyalkanoate Using *Pseudomonas putida* KT2440

**Carolina Ruiz [1], Shane T. Kenny [2], Ramesh Babu P [3], Meg Walsh [2], Tanja Narancic [1,4] and Kevin E. O'Connor [1,4,*]**

1 UCD Earth Institute and School of Biomolecular and Biomedical Science, University College Dublin, Belfield, D04 N2E5, Dublin 4, Ireland; carolina.ruiz@ucdconnect.ie (C.R.); tanja.narancic@ucd.ie (T.N.)
2 Bioplastech Ltd., Nova UCD, Belfield Innovation Park, University College Dublin, Belfield, D04 V2P1, Dublin 4, Ireland; skenny@bioplastech.eu (S.T.K.); meg.walsh@ucdconnect.ie (M.W.)
3 AMBER Centre, CRANN Institute, School of Physics, Trinity College Dublin, Dublin 2, Ireland; BABUP@tcd.ie
4 BEACON—Bioeconomy Research Centre, University College Dublin, D04 N2E5, Belfield, Dublin 4, Ireland
* Correspondence: kevin.oconnor@ucd.ie; Tel.: +353-1-716-2198; Fax: +353-1-716-1183

Received: 30 April 2019; Accepted: 20 May 2019; Published: 21 May 2019

**Abstract:** Waste cooking oil (WCO) is a major pollutant, primarily managed through incineration. The high cell density bioprocess developed here allows for better use of this valuable resource since it allows the conversion of WCO into biodegradable polymer polyhydroxyalkanoate (PHA). WCO was chemically hydrolysed to give rise to a mixture of fatty acids identical to the fatty acid composition of waste cooking oil. A feed strategy was developed to delay the stationary phase, and therefore achieve higher final biomass and biopolymer (PHA) productivity. In fed batch (pulse feeding) experiments *Pseudomonas putida* KT2440 achieved a PHA titre of 58 g/l (36.4% of CDW as PHA), a PHA volumetric productivity of 1.93 g/l/h, a cell density of 159.4 g/l, and a biomass yield of 0.76 g/g with hydrolysed waste cooking oil fatty acids (HWCOFA) as the sole substrate. This is up to 33-fold higher PHA productivity compared to previous reports using saponified palm oil. The polymer (PHA) was sticky and amorphous, most likely due to the long chain monomers acting as internal plasticisers. High cell density cultivation is essential for the majority of industrial processes, and this bioprocess represents an excellent basis for the industrial conversion of WCO into PHA.

**Keywords:** biopolymers; medium chain length polyhydroxyalkanoates (PHA); hydrolysed waste cooking oil; *Pseudomonas putida* KT2440; biocatalysis; bioprocess

---

## 1. Introduction

Petrochemical based plastics have been used in a variety of applications for more than seventy years and have replaced materials like glass and metal due to their high performance, low price, versatility, and durability [1,2]. The worldwide annual production of plastics was 335 million tonnes in 2016, and its production is expected to triple by 2050 [2]. Many of the uses of plastics are short term, and consequently these materials become waste within a short period of time after manufacture and use. The vast majority of petrochemical based plastics are not biologically degraded [3]. Plastic recovery and recycling rates are low [4], and thus millions of tonnes of plastics end up in landfills and in the environment [5–8]. Given the environmental damage caused by non-degradable plastics, there is an urgent need for solutions. Biodegradable plastics can be part of the plethora of solutions to address a complex global challenge. A critical challenge to providing biodegradable plastic solutions is the development of robust processes for biodegradable plastic production.

It is not just the end-of-life of plastics that is a concern for society, but also the origin of the starting materials. Materials of bio-based origin are being sought, as the origins of the current non degradable polymers are finite (fossil based) and depleting [9,10]. Biobased polymers can be produced from renewable resources such as corn dextrose, and many studies are investigating so called second generation (lignocellulose) sources [11–13]. Wastes, such as waste cooking oil, are also potentially interesting starting materials, but no studies have investigated this substrate for high cell density and high PHA productivity.

Polyhydroxyalkanoates (PHAs) are water-insoluble energy storage microbial polyester synthesized by many Gram-positive and Gram-negative bacteria when exposed to a surplus of carbon and generally a limitation of a vital inorganic nutrient (N, P, S, or Mg) [1]. PHAs vary in the composition of the monomer side chain and hydroxyl position, which affects their material properties [14]. The monomer composition, and thus physical properties of PHA, can be tailored by co-feeding different carbon sources in fermentation cultivation systems [15–17]. For example, incorporation of monomers with unsaturated side chains in PHA will increase the melting temperature and decrease the glass transition temperature of the polymer [18]. Furthermore, these unsaturated bonds could be exploited for chemical or enzymatic modifications [19,20] that can render PHAs water soluble or allow their coupling with functional molecules, and therefore broaden their application potential.

High production costs, compared to the traditional petrol-based plastics, remain the major challenge for polyhydroxyalkanoates entry into the plastics market. The use of inexpensive carbon sources, such as waste products, and a highly productive fermentation process could help to overcome the production costs [21–24].

Waste cooking oil (WCO) is a major waste from human food processing and preparation with over 29 million tonnes produced annually around the globe [25]. As WCO contains high levels of fatty acids, it might be suitable substrates for PHA production. It has been widely reported that fatty acids are excellent substrates for PHA accumulation by *Pseudomonas* strains with a PHA composition related to the chain length of the fatty acid supplied [26–30].

Currently, there are a limited number of studies on the use of hydrolysed and saponified cooking oils, or other waste oils, for PHA production [31–34]. Surprisingly, none of these studies have examined the ability of the strains to grow to high cell density on the hydrolysed substrates, nor have they achieved high PHA productivity.

The aim of the current study was to develop a bioprocess to achieve high cell density and high medium chain length PHA (mclPHA) productivity using the fatty acid fraction of hydrolysed waste cooking oil as the sole carbon source and *P. putida* KT2440, a generally recognised as safe (GRAS) and robust organism used in many biotechnological applications [35]. We also examined the properties of the mclPHA polymer accumulated during this bioprocess.

## 2. Results

### 2.1. Fatty Acid Composition of Hydrolysed Waste Cooking Oil (HWCO)

Hydrolysed waste cooking oil (HWCO) was separated into the glycerol and fatty acid fractions using a separation funnel. The fatty acid fraction of hydrolysed waste cooking oil (HWCOFA) contained predominantly oleic and linoleic acid (Figure 1). The high content of C18:2 and C18:1 is in keeping with previously reported contents of fatty acids in waste cooking oils [36–38]. The HWCOFA also had a relatively high content of saturated C16 fatty acid, which makes the hydrolysed oil solid at room temperature. This poses challenges for feeding to a bioreactor and bioavailability in an aqueous growth medium. To address the feeding challenge, the HWCOFA mixture was maintained at a temperature of 40 °C so that it could be poured into the bioreactor.

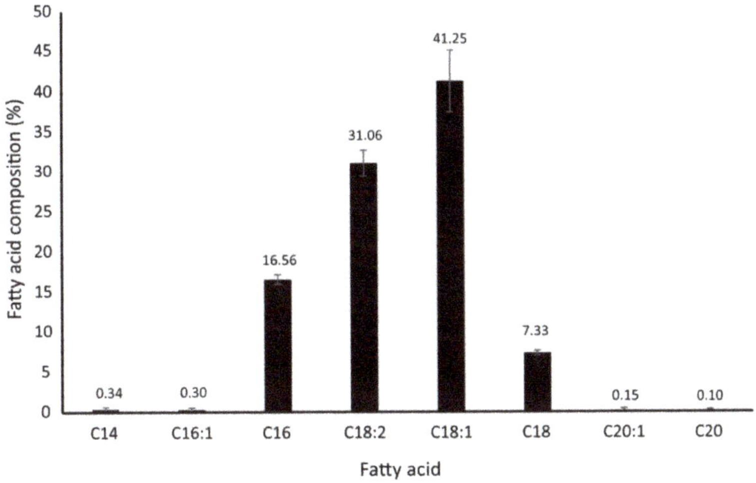


**Figure 1.** Fatty acid composition (expressed as a % of total fatty acids present) of hydrolysed waste cooking oil (HWCO) analysed by GC/MS. C14—myristic acid; C16:1—palmitoleic acid; C16—palmitic acid; C18:2—linolelaidic acid; C18:1—oleic acid; C18—stearic acid; C20:1—gondoic acid; and C20—arachidic acid. The error bar represents standard deviation among three separate hydrolysis experiments.

### 2.2. *Bioprocess Development Using HWCOFA Mixture as the Carbon and Energy Source*

In order to develop a robust growth of *P. putida* KT2440 in a bioreactor using hWCO as carbon substrate, three fed batch strategies were undertaken. All fermentations had the same concentration of 12 g/l HWCOFA at the time of inoculation. Cells were incubated for three hours before any additional HWCOFA was supplied to the bioreactor. The first feed strategy (FS1) is based on the supply of substrate through a pulse feed starting with 12 g of HWCOFA at T3, followed by pulses of the same amount of HWCOFA at T5, T6, and T7 hours (Figure 2). Over the next two hours the pulse feeds were 18 g and 24 g, followed by a 30 g pulse at T11, a 34 g pulse at T12, two 35 g pulses at T13 and T14, and two 40 g pulses at T15 and T16 (Figure 2). Between T17 and T25, 10 pulses of 45 g of HWCOFA were fed. This was followed by 48 g pulses at time 26 and 27 hours, and finally 50 g and 55 g at time 28 and 29 hours, respectively (Figure 2). This feed strategy was based on a feed strategy for nonanoic acid and designed to delay oxygen limitation typically occurring at high cell density growth [39]. The dissolved oxygen in the liquid medium was used as a tool to determine when substrate feeding should occur, and in this strategy, the dissolved oxygen was kept below 5%. A total of 967 g of HWCOFA were supplied over a 30 h incubation, with *P. putida* KT2440 achieving a final CDW of 115 g/l (Figure 3).

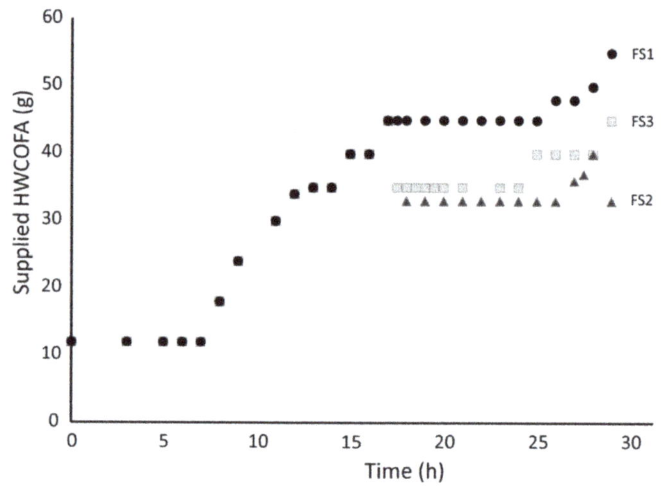


**Figure 2.** Hydrolysed waste cooking oil fatty acids (HWCOFA) pulse feeding strategies. Feed strategies FS2 and FS3 have the same feed profile as FS1 up to T17 hours after, at which time the number of pulses and the amount of HWCOFA supplied was different to FS1. FS1 supplied a total of 967 g of HWCOFA, FS2 804 g of HWCOFA, and FS3 844 g of HWCO, in 30 h fermentation. The starting volume of the culture was 3 l, which increased to 4.2 l for FS1 and FS3 and to 4.1 l for FS2.

We have observed that the growth rate started to slow at 19 hours (Figure 3). Therefore, we designed a second feed strategy (FS2), building on the FS1. The substrate was supplied to the same amount until T17 hours, followed by 9 pulses of 33 g of HWCOFA until T26 hours, and pulses of 36 g, 37 g, 40 g, and 33 g at T27, T27.5, T28, and T39 (Figure 2). FS2 supplied a total of 804 g of HWCOFA, achieving a final biomass of 145 g/l with a delayed onset of the stationary phase observed (Figure 3). The final strategy, FS3, was also an adaptation of FS1 with the same amount of HWCOFA supplied to the bioreactor up to 17 hours, followed by nine 35 g pulses between T17.5 and T23, four pulses of 40 g between T24 and T28, and a final 45 g pulse at T29 (Figure 2). The total HWCOFA supplied to the fermentation medium was 844 g, which resulted in a final CDW of 159 g/l (Figure 3). PHA accumulation was detected after approximately 10 hours of incubation and linearly increased for longer in FS2 and FS3 (Figure 4). The highest substrate to biomass yield of was achieved with FS3 resulting in 1.5-fold improvement compared to FS1, while substrate to PHA yield increased 1.9-fold, with FS3 compared to FS1 (Table 1). Interestingly, the PHA content of cells decreased towards the end of the bioprocess with FS2 and FS3 (Figure 4), but without the corresponding drop in biomass (Figure 2).

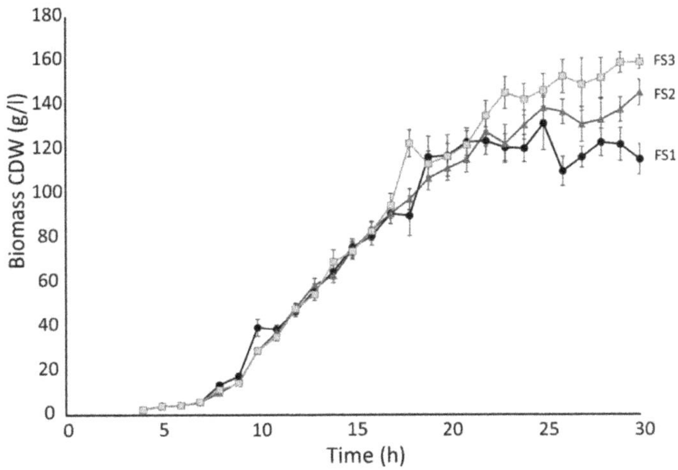


**Figure 3.** Growth of *P. putida* KT2440 on hydrolysed waste cooking oil fatty acids (HWCOFA) using three different feed strategies: FS1, FS2, and FS3. Biomass is represented as cell dry weight (CDW; g/l). Data are the average of three independent biological replicates and the error bars represent standard deviation among these.

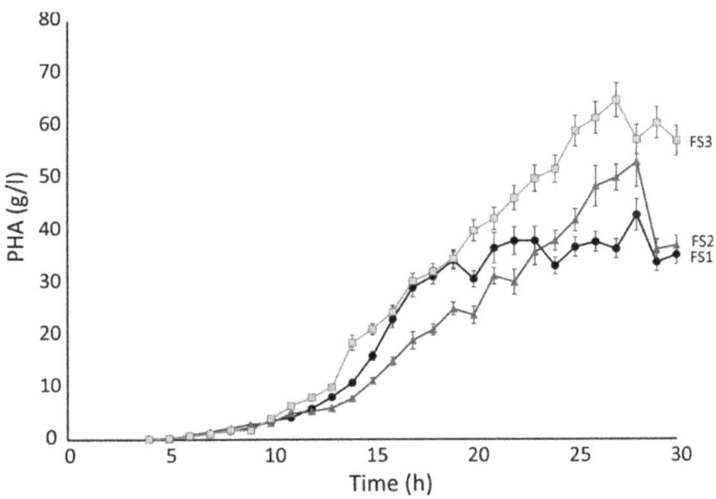


**Figure 4.** The dynamics of PHA (g/l) accumulation by *P. putida* KT2440 resulting from feed strategies FS1, FS2, and FS3. The error bars represent standard deviation among three independent biological replicates.

**Table 1.** Growth data for *P. putida* KT2440 using different feeding strategies.

| Substrate | Feed Strategy | Initial Volume (l) | Total Substrate Used (g) | Final CDW (g/l) | Final Volume (l) | Total Biomass (g) |
|---|---|---|---|---|---|---|
| HWCOFAs * | 1 | 3 | 967 | 115.1 | 4.2 | 483.4 |
| | **Final PHA (%CDW)** | **Final PHA (g/l)** | **PHA productivity (g/l/h)** | **Total PHA produced (g)** | **Biomass Yield (g/g)** | **PHA Yield (g/g)** |
| | 30.6 | 35.2 | 1.17 | 147.9 | 0.50 | 0.15 |
| HWCOFAs | **Feed strategy** | **Initial Volume (l)** | **Total substrate used (g)** | **Final CDW (g/l)** | **Final Volume (l)** | **Total Biomass (g)** |
| | 2 | 3 | 804 | 145.2 | 4.1 | 595.3 |
| | **Final PHA (%CDW)** | **Final PHA (g/l)** | **PHA productivity (g/l/h)** | **Total PHA produced (g)** | **Biomass Yield (g/g)** | **PHA Yield (g/g)** |
| | 25.5 | 37.0 | 1.23 | 151.8 | 0.74 | 0.19 |
| HWCOFAs | **Feed strategy** | **Initial Volume (l)** | **Total substrate used (g)** | **Final CDW (g/l)** | **Final Volume (l)** | **Total Biomass (g)** |
| | 3 | 3 | 881 | 159.4 | 4.2 | 669.5 |
| | **Final PHA (%CDW)** | **Final PHA (g/l)** | **PHA productivity (g/l/h)** | **Total PHA produced (g)** | **Biomass Yield (g/g)** | **PHA Yield (g/g)** |
| | 36.4 | 58.0 | 1.93 | 243.7 | 0.76 | 0.28 |

* HWCOFAs: hydrolysed waste cooking oil fatty acids. Data are the average of three independent biological replicates (SD < 5%).

### 2.3. Polymer Analysis

The specific industrial application of PHAs is determined by the mechanical and thermal properties of the biopolymer. These characteristics are affected by the monomer composition, which varies according to the metabolic machinery of the organisms and the substrate supplied [40,41].

The polymer accumulated by *P. putida* KT2440 using HWCOFA is sticky, waxy, and has low molecular weight (Table 2). It contained 42 mol% of (*R*)-3-hydroxydecanoic acid, 39 mol% of (*R*)-3hydroxyoctanoic acid, 8 mol% of (*R*)-3-hydroxydodecanoic acid, 6 mol% of (*R*)-3-hydroxyundodecanoic acid, and 5 mol% of (*R*)-3-hydroxyhexanoic acid. The polydispersity value of just under 2 is typical for mclPHAs (Table 2). Thermal analysis showed a similar low glass transition temperature for both polymers (Table 2). These properties indicate the polymer could be useful for adhesive applications.

**Table 2.** Properties of PHA polymer extracted from *P. putida* KT2440 grown on hydrolysed waste cooking oil fatty acids (HWCOFAs).

| Substrate Conditions | Td (°C) | Tg (°C) | Tm (°C) | Mn | Mw |
|---|---|---|---|---|---|
| **HWCOFAs** | 270.61 ± 1.6 | -56.1 ± 0.5 | 20.7 ± 0.5 | 22954 ± 975 | 45317 ± 62 |

Td—thermal degradation temperature; Tg—glass transmission temperature; Tm—melting temperature; Mn—number; Mw—molecular weight.

## 3. Discussion

We have developed a successful bioprocess for the high cell density conversion of HWCOFA into mclPHA with high PHA volumetric productivity (Table 1). WCO represents a major pollutant due to the poor end-of-life management [42]. The dominant route for the WCO management is incineration, which leads to the loss of the valuable resource [43]. The bioprocess reported here contributes to the end-of-life management through valorisation of this waste material and production of a biodegradable polymer.

While the use of modelling informs the strategy to optimise processes and reduces unnecessary experimentation [44], the building of the model depends upon the accurate measurement of variables, such as substrate concentration. Long-chain length fatty acids such as those found in HWCO are insoluble in water [45]. The hydrophobicity of the fatty acids results in a lack of uniformity in their dispersion in an aqueous medium and the sticking of fatty acids to the wall of the bioreactor. Thus, it was not possible to accurately measure substrate utilisation and a mathematical model of the process could not be constructed. Therefore, an empirical approach was taken in developing a bioprocess for conversion of HWCOFA to mclPHA. Oxygen limitation is a typical reason for termination of aerobic high cell density fermentations, causing the onset of the stationary phase [39]. Oxygen uptake rate is proportional to the cell generation rate, and therefore a gradual decrease in growth rate by decaying the amount of substrate fed should result in high cell density while avoiding the negative physiological effects of growth halt. To initiate the fed batch fermentation, 12 g/l of HWCOFA was used as the starting concentration, as this is known to be below the inhibitory concentration of fatty acid substrates [46]. The fed batch process began with the first feed of HWCOFAs 3 h after inoculation. The rate of substrate feeding then was varied by increasing amounts of the fed substrate, followed by constant feed of the substrate throughout the exponential to early stationary phase (Figure 2). Decaying feeds in FS2 and FS3 caused a delayed onset of stationary phase, and therefore higher final biomass (Figure 3). Decaying feeds also resulted in increased PHA accumulation, likely due to preventing the accumulation of the inhibitory level of fatty acids.

Another challenge we encountered when developing this bioprocess was foaming. The same issue was described when crude fatty acids were supplied as a carbon substrate in bioreactors at concentrations greater than 10 g/l [31]. Thus the implementation of this technology at scale will require good foam breakage and dispersion.

High PHA productivity when fatty acids are used as a substrate was reported earlier. Lee and co-workers observed *Aeromonas hydrophila* cells with 45.2% of the CDW as PHA accumulation, and a PHA productivity of 1.01 g/l/h with lauric acid and oleic acid [47]. *P. putida* KT2442 grown on oleic acid and *P. putida* KT2440 grown on nonanoic acid have achieved productivities at 1.9 g/l/h and 2.3 g/l/h, respectively [39,47]. In this work, when HWCOFA was used as a substrate, the PHA volumetric productivity was 1.9 g/l/h (Table 1). Thus, low cost HWCOFA show as much potential for PHA production as more expensive virgin fatty acids.

While a number of studies investigated the production of mclPHAs from vegetable oils, waste cooking oils, and saponified waste palm oil [48–51], this is the first study to use HWCOFA for mclPHA production, with high cell density and high PHA productivity in a bioreactor. For example, a mixture of glucose and 2% free fatty acids was used for the growth and PHA accumulation by *P. aeruginosa* 47T2 in shaken flasks [52]. While achieved PHA levels were comparable between this and our study, the biomass (7.9 g/l CDW) reported by Haba et al. [52] is 18.7-fold lower than the biomass achieved in the present study. Similarly, the ability of *P. oleovorans*, *P. resivorans*, and *P. putida* to produce mclPHA from tallow was demonstrated only at low cell density [53]. Conversion of saponified waste palm oil as the sole source of carbon and energy into PHA in a bioreactor by *Pseudomonas sp.* Gl01 was reported by Możejko and Ciesielski [51]. However, our bioprocess based on FS3 achieves 33-fold higher PHA productivity compared to productivity reported by Możejko and Ciesielski [51].

The vast majority of studies report on the need for inorganic nutrient limitation in order to observe PHA accumulation in *Pseudomonas* and other strains [24,54,55]. In the current study, *P. putida* KT2440 accumulated PHA early in the growth cycle when no inorganic nutrient limitation occurred. The fatty acids present in HWCO feed directly into beta-oxidation. The resultant intermediates are PHA precursors. Given the high energy state of cells as a result of growth on long chain fatty acids and the presence of precursors in the cell it is not surprising that we observed PHA accumulation, as the conditions for accumulation of PHA were ideal.

PHA accumulated by bacteria supplied with vegetable oil derived free fatty acids and animal fat derived free fatty acids display liquid properties at room temperature similar to the polymers arising from the supply of HWCOFAs to bacteria in the current study [21,56]. These polymer properties can be explained by the high degree of disorder and by the long chain monomers acting as internal plasticisers [30]. However, the molecular weight (Mw) of PHA polymers accumulated by *Pseudomonas species* supplied with other fatty acid substrates, such as oleic acid, lauric acid [57], and mystiric acid [34], were between 2 and 4.2-fold higher than those seen in our study.

In conclusion, we have developed a high cell density cultivation with the highest known PHA productivity bioprocess using WCO as a source of waste fatty acids. The PHA productivity is similar to that achieved with pure single source fatty acids [28,34,47] and shows great promise for process scale up and polymer product development.

## 4. Materials and Methods

### 4.1. Bacterial Growth Medium & Strain Maintenance

Minimal Salt Medium (MSM) was used as the growth medium for all strains in shaken flask experiments. This medium contains (per litre): 9 g $Na_2HPO_4 x 12 x H_20$, 1.5 g $KH_2PO_4$, 1 g $NH_4Cl$, 200 μg $MgSO_4$, and 1000 μL trace elements solution. The trace elements solution contained (per litre in 1M HCl): 4 g $ZnSO_{4}x7H_20$, 10 g $FeSO_{4.}\ x\ 7H_20$, 1 g $CuCl_{2}x2H_20$, 1 g $MnCl_{2}x4H_20$, 1 g $Na_2B_4O7_{x}10H_20$, 0.2 g $NiCl_{2}x6H_20$, and 0.3g $Na_2MoO_{4}x2H_20$. *Pseudomonas putida* KT2440 [ATCC® 47054™] was grown on *Pseudomonas* isolation agar (PIA) (Fluka analytical).

### 4.2. Waste Cooking Oil (WCO) Hydrolysis

The WCO was supplied by Frylite® (Dublin, Ireland). A measure of 100 g of WCO was hydrolysed using 100 mL of 6 M NaOH. The mixture was heated at 60 °C for 90 min. The resultant saponified fatty acid mixture was precipitated through the addition of 400 mL of 6 M HCl. The mixture was decanted into a separating funnel and the fatty acid phase separated from the aqueous phase, containing salts and glycerol. The fatty acid phase was washed with approximately 1000 mL of distilled water, filtered using a vacuum system (Fisher FB59037 Range QL100, Dublin, Ireland), and supplied as the sole carbon and energy substrate in fermentations.

### 4.3. Fermentation Conditions

A single colony of *P. putida* KT2440 from a PIA plate was inoculated into 50 mL Minimal Salt Media (MSM) in flask experiments, as previously described [58]. Flasks were supplemented with 3 g/l of technical oleic acid (Sigma Aldrich, Dublin, Ireland). Flasks were incubated at 30 °C, shaking at 200 rpm for 16–18 h. Four 50 mL cultures of *P. putida* KT2440 overnights were prepared for inoculation into a fermenter.

Fermentations were performed in a Biostat B+ bioreactor with a 5 litre working volume (Sartorius). MSM was used as the base media for all fermentations. The fermentations had an initial volume of 3 litres, with an initial agitation of 500 rpm. The temperature was maintained at 30 °C and pH was controlled at 6.9 +/− 0.1 by the addition of 20% $NH_4OH$ solution or 15% (v/v) $H_2SO_4$. The $NH_4OH$ also acted as a nitrogen source. Foaming was controlled by the addition of antifoam solution (polypropylene glycol P2000, Sigma). Dissolved oxygen (DO) was set at 20% of saturation and was increased by increasing agitation. $O_2$ was supplied as air at a constant flow rate of 5 litres per minute. To start, the fermentation substrate was supplied at a concentration of 12 g/l. Three pulse feeding strategies for HWCOFA were undertaken and are described in the results section.

### 4.4. Analysis of Fatty Acids in Hydrolysed Waste Cooking Oil

Fatty acids generated by hydrolysis of WCO and virgin plant oil standards (Sigma Aldrich) were derivatised with *N*-Methyl-N-(trimethylsilyl)trifluoroacetamide (TMS). A measure of 2 mL of chloroform was placed in a gas chromatography vial. Then, 1 µL of the substrate and 20 µL of TMS were added. The vial was incubated at 70 °C for 30 min. The fatty acids were then analysed using an Agilent 6890N gas chromatograph (GC) (Cork, Ireland) fitted with a 5973 series inert mass spectrophotometer (MS). A HP-5 column (12 m × 0.2 mm × 0.33 µm; Hewlett Packard) was used with an oven method of 50 °C for 3 min, increasing by 10 °C /min to 250 °C and holding for 1 min. A 10:12 split was used with helium as the carrier gas and an inlet temperature of 250 °C.

### 4.5. PHA Content and Monomer Composition Determination

Determination of the PHA content of cells and monomer composition of PHA was determined by subjecting lyophilised cells to acidic methanolysis [59,60]. Dried cells were weighed (5 to 10 mg) and suspended in 2 mL of acidified methanol (15% $H_2SO_4$, v/v) and 2 mL of chloroform, containing 6 mg/l benzoate methyl ester as an internal standard. The solution was placed in 15 ml Pyrex test tubes, sealed, and incubated at 100 °C for 3 h. The tubes were then placed on ice for 1 min. A meausre of 1 mL of water was added to each tube, and the solution was mixed by vigorous vortexing for 1 min. The phases were allowed to separate, and the organic phase was removed and passed through a filter before further analysis. The 3-hydroxyalkanoic acid methyl esters were analysed by gas chromatography (GC) using an Agilent 6890N chromatograph equipped with a HP Innowax column (30 m × 0.25 mm × 0.5 µm) and a flame ionisation detector (FID). An oven ramp cycle was employed as follows: 120 °C for 5 min, increasing by 3 °C/min to 180 °C, and at 180 °C for 10 min. A 20:1 split was used with helium as the carrier gas and an inlet temperature of 250 °C. Commercially available 3-hydroxyalkanoic acids (Bioplastech Ltd Dublin Ireland) were methylated as described above for PHA isolated from *P. putida* KT2440, and they were used as standards to identify individual PHA monomers. The unsaturated monomers were detected using the programme described for the analysis of fatty acids.

### *4.6. Nutrient and Biomass Analysis*

Samples were taken at 1 or 2 h intervals during the fermentations. Two 2 mL samples were taken and centrifuged at 17.960× *g* for 3 min. The supernatant was decanted into a separate tube, and the cell pellets and supernatant were frozen at –80 °C. Cell dry weight (CDW) was determined by first freezing the cell pellet and then lyophilising (freeze-drying) the cell pellet at –80 °C overnight and subsequently weighing the cells. For cultures grown in a 50 mL flask, cell suspensions were centrifuged at 3,220× *g* for 10 min, and 2 mL of supernatant was retained and the remainder discarded. Pellets were resuspended in 1 mL of DI water and then transferred into Eppendorf tubes. These were then centrifuged at 17.960× *g* for 3 min; the supernatant was discarded and the pellets treated as the samples from the bioreactor. The concentration of nitrogen in the supernatant was determined using the method described by Scheiner [61]. The soluble inorganic phosphate concentration was determined using the USA EPA colorimetric method (USEPA, 1978).

### *4.7. Polymer Isolation*

Cells were harvested from the bioreactor in a Sorvall centrifuge (Fisher Scientific, Dublin, Ireland) at 25,040 g. Harvested cells were frozen at −80 °C for 24 h and then lyophilised (Labconco, Fisher Scientific). PHA was isolated from freeze-dried cells using room temperature acetone. This involved the stirring of 10 g of cells suspended in 100 mL acetone for 24 h. The mixture was allowed to settle, and the supernatant was filtered using a 0.2 µm PTFE filter. Acetone containing PHA was then subjected to rotary evaporation under vacuum until approximately 90 mL of acetone had been recovered. The polymer was precipitated using 2 vol of a wash solution consisting of 35% methanol, 35% ethanol, and 30% distilled water [62]. The supernatant was then decanted, and the precipitated PHA was allowed to dry before further analysis.

### *4.8. PHA Characterisation*

#### 4.8.1. Gel Permeation Chromatography (GPC)

The average molecular weight (Mw), the molecular number (Mn), and the polydispersity index (PD) of the polymer were measured by GPC using PL gel 5 mm mixed-C +PL gel column (Perkin-Elmer, Dublin, Ireland)) with the PELV 290 UV-vis detector set at 254 nm. Spectroscopic grade chloroform was used as the eluent flow rate of 1.0 mL/min. A sample concentration of 1% (w/v) and injection volumes of 500 µL were used. A molecular weight calibration curve was generated with polystyrene standards with low polydispersity [63].

#### 4.8.2. Differential Scanning Calorimetry (DSC)

The polymer was analysed by DSC (Perkin-Elmer, Dublin,) with a Perkin–Elmer Pyris Diamond calorimeter calibrated to Indium standards to determine the glass transition temperature (Tg), melting temperature (Tm), and degradation temperature (Td). The samples were encapsulated in hermetically sealed aluminium pans and heated from −70 °C to 100 °C at a rate of 10 °C/min. To determine the glass transition temperature (Tg) the samples were held at 100 °C for 1 min and rapidly quenched to −70 °C. The samples were then reheated from −70 to 100 °C at 10 °C/min to determine the melting temperature (Tm) and glass transition temperature (Tg). Finally, the samples were heated to 350 °C at a rate of 10 °C/min to determine the thermal destruction temperature (Td) [63].

**Author Contributions:** Conceptualization, S.T.K. and K.E.O.; Funding acquisition, K.E.O.; Investigation, C.R., R.B.P., and M.W.; Methodology, C.R., S.T.K., and T.N.; Supervision, S.T.K., T.N., and K.E.O.; Validation, K.E.O.; Writing—original draft, C.R.; Writing—review & editing, S.T.K., T.N. and K.E.O.

**Acknowledgments:** Carolina Ruiz and this work were funded by Coordenação de Aperfeiçoamento de Pessoal de Nível Superior (CAPES), Brazil.

**Conflicts of Interest:** The authors declare no conflict of interest.

## References


1. Keshavarz, T.; Roy, I. Polyhydroxyalkanoates: bioplastics with a green agenda. *Curr. Opin. Microbiol.* **2010**, *13*, 321–326. [CrossRef]
2. Plastics Europe. Plastics—The Facts 2017; 2017. Available online: https://www.plasticseurope.org/en/resources/publications/274-plastics-facts-2017 (accessed on 23 April 2019).
3. Narancic, T.; O'Connor, K.E. Plastic waste as a global challenge: are biodegradable plastics the answer to the plastic waste problem. *Microbiology* **2019**, *165*, 129–137. [CrossRef]
4. System Initiative on Environment and Natural Resource Security. The New Plastics Economy: Catalysing action. World Economic Forum: 2017. Available online: https://www.ellenmacarthurfoundation.org/publications/new-plastics-economy-catalysing-action.
5. Hanke, G. *Marine Beach Litter in Europe—Top Items*; Joint Research Centre, European Commission, 2016; Available online: https://mcc.jrc.ec.europa.eu/documents/Marine_Litter/MarineLitterTOPitems_final_24.1.2017.pdf.
6. Rochman, C.M.; Browne, M.A.; Halpern, B.S.; Hentschel, B.T.; Hoh, E.; Karapanagioti, H.K.; Rios-Mendoza, L.M.; Takada, H.; Teh, S.; Thompson, R.C. Classify plastic waste as hazardous. *Nature* **2013**, *494*, 169–171. [CrossRef] [PubMed]
7. Rochman, C.M.; Hoh, E.; Kurobe, T.; Teh, S.J. Ingested plastic transfers hazardous chemicals to fish and induces hepatic stress. *Sci. Rep.* **2013**, *3*, 1–7. [CrossRef] [PubMed]
8. Wilcox, C.; Van Sebille, E.; Hardesty, B.D. Threat of plastic pollution to seabirds is global, pervasive, and increasing. *Proc. Natl. Acad. Sci. USA* **2015**, *112*, 11899–11904. [CrossRef] [PubMed]
9. Mecking, S. Nature or petrochemistry? Biologically degradable materials. *Angew. Chem. Int. Ed.* **2004**, *43*, 1078–1085. [CrossRef]
10. Babu, R.P.; O'Connor, K.; Seeram, R. Current progress on bio-based polymers and their future trends. *Prog. Biomater.* **2013**, *2*, 8. [CrossRef]
11. Mohanty, A.K.; Misra, M.; Drzal, L.T. Sustainable bio-composites from renewable resources: Opportunities and challenges in the green materials world. *J. Polym. Environ.* **2002**, *10*, 19–26. [CrossRef]
12. Harmsen, P.F.H.; Hackmann, M.M.; Bos, H.L. Green building blocks for bio-based plastics. *Biofuel Bioprod. Bior.* **2014**, *8*, 306–324. [CrossRef]
13. Isikgor, F.H.; Becer, C.R. Lignocellulosic biomass: a sustainable platform for the production of bio-based chemicals and polymers. *Polym. Chem.* **2015**, *6*, 4497–4559. [CrossRef]
14. Sudesh, K.; Abe, H.; Doi, Y. Synthesis, structure and properties of polyhydroxyalkanoates: Biological polyesters. *Prog. Polym. Sci.* **2000**, *25*, 1503–1555. [CrossRef]
15. Malacara, C.F.P.; Romero, A.G.; Ponce, M.M.; Marenco, T.C. Approaches for the synthesis of tailor-made polyhydroxyalkanoates. In *Microbial Factories: Biodiversity, Biopolymers, Bioactive Molecules: Volume 2*; Kalia, V.C., Ed.; Springer India: New Delhi, India; pp. 11–28. [CrossRef]
16. Kang, H.O.; Chung, C.W.; Kim, H.W.; Kim, Y.B.; Rhee, Y.H. Cometabolic biosynthesis of copolyesters consisting of 3-hydroxyvalerate and medium-chain-length 3-hydroxyalkanoates by Pseudomonas sp DSY-82. *Anton. Leeuw. Int. J. G.* **2001**, *80*, 185–191. [CrossRef]
17. Lenz, R.W.; Kim, Y.B.; Fuller, R.C. Production of unusual bacterial polyesters by Pseudomonas oleovorans through cometabolism. *Fems Microbiol. Lett.* **1992**, *103*, 207–214. [CrossRef]
18. Ashby, R.D.; Foglia, T.A. Poly(hydroxyalkanoate) biosynthesis from triglyceride substrates. *Appl. Microbiol. Biot.* **1998**, *49*, 431–437. [CrossRef]
19. Li, Z.B.; Loh, X.J. Water soluble polyhydroxyalkanoates: future materials for therapeutic applications. *Chem. Soc. Rev.* **2015**, *44*, 2865–2879. [CrossRef] [PubMed]
20. Vastano, M.; Pellis, A.; Immirzi, B.; Dal Poggetto, G.; Malinconico, M.; Sannia, G.; Guebitz, G.M.; Pezzella, C. Enzymatic production of clickable and PEGylated recombinant polyhydroxyalkanoates. *Green. Chem.* **2017**, *19*, 5494–5504. [CrossRef]
21. Walsh, M.; O'Connor, K.; Babu, R.; Woods, T.; Kenny, S. Plant oils and products of their hydrolysis as substrates for polyhydroxyalkanoate synthesis. *Chem. Biochem. Eng. Q.* **2015**, *29*, 123–133. [CrossRef]
22. Sabapathy, P.C.; Devaraj, S.; Kathirvel, P. Parthenium hysterophorus: low cost substrate for the production of polyhydroxyalkanoates. *Curr. Sci.* **2017**, *112*, 2106–2111. [CrossRef]

23. Poblete-Castro, I.; Binger, D.; Oehlert, R.; Rohde, M. Comparison of mcl-Poly(3-hydroxyalkanoates) synthesis by different Pseudomonas putida strains from crude glycerol: citrate accumulates at high titer under PHA-producing conditions. *BMC Biotechnol.* **2014**, 14. [CrossRef]
24. Ward, P.G.; Goff, M.; Donner, M.; Kaminsky, W.; O'Connor, K.E. A two step chemo-biotechnological conversion of polystyrene to a biodegradable thermoplastic. *Environ. Sci. Technol.* **2006**, *40*, 2433–2437. [CrossRef]
25. Maddikeri, G.L.; Pandit, A.B.; Gogate, P.R. Intensification approaches for biodiesel synthesis from waste cooking oil: A review. *Ind. Eng. Chem. Res.* **2012**, *51*, 14610–14628. [CrossRef]
26. Sanchez, R.J.; Schripsema, J.; da Silva, L.F.; Taciro, M.K.; Pradella, J.G.C.; Gomez, J.G.C. Medium-chain-length polyhydroxyalkanoic acids (PHA(mcl)) produced by Pseudomonas putida IPT 046 from renewable sources. *Eur. Polym. J.* **2003**, *39*, 1385–1394. [CrossRef]
27. Le Meur, S.; Zinn, M.; Egli, T.; Thony-Meyer, L.; Ren, Q. Production of medium-chain-length polyhydroxyalkanoates by sequential feeding of xylose and octanoic acid in engineered Pseudomonas putida KT2440. *BMC Biotechnol.* **2012**, *12*, 1–12. [CrossRef]
28. Sun, Z.; Ramsay, J.A.; Guay, M.; Ramsay, B. Increasing the yield of MCL-PHA from nonanoic acid by co-feeding glucose during the PHA accumulation stage in two-stage fed-batch fermentations of Pseudomonas putida KT2440. *J. Biotechnol.* **2007**, *132*, 280–282. [CrossRef]
29. Kim, B.S. Production of medium chain length polyhydroxyalkanoates by fed-batch culture of Pseudomonas oleovorans. *Biotechnol. Lett.* **2002**, *24*, 125–130. [CrossRef]
30. Kellerhals, M.B.; Kessler, B.; Witholt, B.; Tchouboukov, A.; Brandl, H. Renewable long-chain fatty acids for production of biodegradable medium-chain-length polyhydroxyalkanoates (mcl-PHAs) at laboratory and pilot plant scales. *Macromolecules* **2000**, *33*, 4690–4698. [CrossRef]
31. Annuar, M.S.M.; Tan, I.K.P.; Ibrahim, S.; Ramachandran, K.B. Production of medium-chain-length poly(3-hydroxyalkanoates) from crude fatty acids mixture by Pseudomonas putida. *Food Bioprod Process* **2007**, *85*, 104–119. [CrossRef]
32. Fernández, D.; Rodríguez, E.; Bassas, M.; Viñas, M.; Solanas, A.M.; Llorens, J.; Marqués, A.M.; Manresa, A. Agro-industrial oily wastes as substrates for PHA production by the new strain Pseudomonas aeruginosa NCIB 40045: Effect of culture conditions. *Biochem. Eng. J.* **2005**, *26*, 159–167. [CrossRef]
33. Mozejko, J.; Wilke, A.; Przybylek, G.; Ciesielski, S. Mcl-PHAs produced by Pseudomonas sp Gl01 using fed-batch cultivation with waste rapeseed oil as carbon source. *J. Microbiol. Biotechn.* **2012**, *22*, 371–377. [CrossRef]
34. Tan, I.K.P.; Kumar, K.S.; Theanmalar, M.; Gan, S.N.; Gordon, B. Saponified palm kernel oil and its major free fatty acids as carbon substrates for the production of polyhydroxyalkanoates in Pseudomonas putida PGA1. *Appl. Microbiol. Biot.* **1997**, *47*, 207–211. [CrossRef]
35. Belda, E.; van Heck, R.G.A.; Lopez-Sanchez, M.J.; Cruveiller, S.; Barbe, V.; Fraser, C.; Klenk, H.P.; Petersen, J.; Morgat, A.; Nikel, P.I.; et al. The revisited genome of *Pseudomonas putida* KT2440 enlightens its value as a robust metabolic chassis. *Environ. Microbiol.* **2016**, *18*, 3403–3424. [CrossRef] [PubMed]
36. Leung, D.Y.C.; Guo, Y. Transesterification of neat and used frying oil: Optimization for biodiesel production. *Fuel Process Technol.* **2006**, *87*, 883–890. [CrossRef]
37. Cruz, M.V.; Sarraguca, M.C.; Freitas, F.; Lopes, J.A.; Reis, M.A.M. Online monitoring of P(3HB) produced from used cooking oil with near-infrared spectroscopy. *J. Biotechnol.* **2015**, *194*, 1–9. [CrossRef]
38. Martino, L.; Cruz, M.V.; Scoma, A.; Freitas, F.; Bertin, L.; Scandola, M.; Reis, M.A.M. Recovery of amorphous polyhydroxybutyrate granules from Cupriavidus necator cells grown on used cooking oil. *Int. J. Biol. Macromol.* **2014**, *71*, 117–123. [CrossRef]
39. Maclean, H.; Sun, Z.Y.; Ramsay, J.; Ramsay, B. Decaying exponential feeding of nonanoic acid for the production of medium-chain-length poly(3-hydroxyalkanoates) by Pseudomonas putida KT2440. *Can. J. Chem.* **2008**, *86*, 564–569. [CrossRef]
40. Lemos, P.C.; Serafim, L.S.; Reis, M.A.M. Synthesis of polyhydroxyalkanoates from different short-chain fatty acids by mixed cultures submitted to aerobic dynamic feeding. *J. Biotechnol.* **2006**, *122*, 226–238. [CrossRef] [PubMed]
41. Zhila, N.; Shishatskaya, E. Properties of PHA bi-, ter-, and quarter-polymers containing 4-hydroxybutyrate monomer units. *Int. J. Biol. Macromol.* **2018**, *111*, 1019–1026. [CrossRef]

42. Nantha Gopal, K.; Pal, A.; Sharma, S.; Samanchi, C.; Sathyanarayanan, K.; Elango, T. Investigation of emissions and combustion characteristics of a CI engine fueled with waste cooking oil methyl ester and diesel blends. *Alex. Eng. J.* **2014**, *53*, 281–287. [CrossRef]
43. Taniguchi, I.; Kagotani, K.; Kimura, Y. Microbial production of poly(hydroxyalkanoate)s from waste edible oils. *Green Chem.* **2003**, *5*, 545–548. [CrossRef]
44. Giridhar, R.; Srivastava, A.K. Fed-batch sorbose fermentation using pulse and multiple feeding strategies for productivity improvement. *Biotechnol. Bioprocess Eng.* **2000**, *5*, 340–344. [CrossRef]
45. Lie, E.; Molin, G. Hydrolysis and esterification with immobilized lipase on hydrophobic and hydrophilic zeolites. *J. Chem. Technol. Biot.* **1991**, *50*, 549–553. [CrossRef]
46. Lee, S.Y.; Wong, H.H.; Choi, J.I.; Lee, S.H.; Lee, S.C.; Han, C.S. Production of medium-chain-length polyhydroxyalkanoates by high-cell-density cultivation of Pseudomonas putida under phosphorus limitation. *Biotechnol. Bioeng.* **2000**, *68*, 466–470. [CrossRef]
47. Lee, S.H.; Oh, D.H.; Ahn, W.S.; Lee, Y.; Choi, J.I.; Lee, S.Y. Production of poly(3-hydroxybutyrate-co-3-hydroxyhexanoate) by high-cell-density cultivation of Aeromonas hydrophila. *Biotechnol. Bioeng.* **2000**, *67*, 240–244. [CrossRef]
48. Marsudi, S.; Unno, H.; Hori, K. Palm oil utilization for the simultaneous production of polyhydroxyalkanoates and rhamnolipids by Pseudomonas aeruginosa. *Appl. Microbiol. Biot.* **2008**, *78*, 955–961. [CrossRef] [PubMed]
49. Verlinden, R.A.J.; Hill, D.J.; Kenward, M.A.; Williams, C.D.; Piotrowska-Seget, Z.; Radecka, I.K. Production of polyhydroxyalkanoates from waste frying oil by Cupriavidus necator. *AMB Exp.* **2011**, *1*, 1–8. [CrossRef] [PubMed]
50. Yun, H.S.; Kim, D.Y.; Chung, C.W.; Kim, H.W.; Yang, K.Y.; Rhee, Y.H. Characterization of a tacky poly(3-hydroxyalkanoate) produced by Pseudomonas chlororaphis HS21 from palm kernel oil. *J. Microbiol. Biotechn.* **2003**, *13*, 64–69.
51. Mozejko, J.; Ciesielski, S. Saponified waste palm oil as an attractive renewable resource for mcl-polyhydroxyalkanoate synthesis. *J. Biosci. Bioeng.* **2013**, *116*, 485–492. [CrossRef] [PubMed]
52. Haba, E.; Vidal-Mas, J.; Bassas, M.; Espuny, M.J.; Llorens, J.; Manresa, A. Poly 3-(hydroxyalkanoates) produced from oily substrates by Pseudomonas aeruginosa 47T2 (NCBIM 40044): Effect of nutrients and incubation temperature on polymer composition. *Biochem. Eng. J.* **2007**, *35*, 99–106. [CrossRef]
53. Cromwick, A.M.; Foglia, T.; Lenz, R.W. The microbial production of poly(hydroxyalkanoates) from tallow. *Appl. Microbiol. Biot.* **1996**, *46*, 464–469. [CrossRef]
54. Tobin, K.M.; O'Connor, K.E. Polyhydroxyalkanoate accumulating diversity of Pseudomonas species utilising aromatic hydrocarbons. *Fems Microbiol. Lett.* **2005**, *253*, 111–118. [CrossRef] [PubMed]
55. Valentino, F.; Karabegouic, L.; Majone, M.; Morgan-Sagastume, F.; Werker, A. Polyhydroxyalkanoate (PHA) storage within a mixed-culture biomass with simultaneous growth as a function of accumulation substrate nitrogen and phosphorus levels. *Water Res.* **2015**, *77*, 49–63. [CrossRef] [PubMed]
56. Muhr, A.; Rechberger, E.M.; Salerno, A.; Reiterer, A.; Malli, K.; Strohmeier, K.; Schober, S.; Mittelbach, M.; Koller, M. Novel description of mcl-PHA biosynthesis by Pseudomonas chlororaphis from animal-derived waste. *J. Biotechnol.* **2013**, *165*, 45–51. [CrossRef] [PubMed]
57. Gumel, A.M.; Annuar, M.S.M.; Heidelberg, T. Growth kinetics, effect of carbon substrate in biosynthesis of mcl-PHA by Pseudomonas putida Bet001. *Braz. J. Microbiol.* **2014**, *45*, 427–438. [CrossRef]
58. Schlegel, H.G.; Kaltwasser, H.; Gottschalk, G. A submersion method for culture of hydrogen-oxidizing bacteria: growth physiological studies. *Arch. Mikrobiol.* **1961**, *38*, 209–222. [CrossRef]
59. Brandl, H.; Gross, R.A.; Lenz, R.W.; Fuller, R.C. Pseudomonas oleovorans as a source of poly(beta-hydroxyalkanoates) for potential applications as biodegradable polyesters. *Appl. Environ. Microb.* **1988**, *54*, 1977–1982.
60. Lageveen, R.G.; Huisman, G.W.; Preusting, H.; Ketelaar, P.; Eggink, G.; Witholt, B. Formation of polyesters by Pseudomonas oleovorans—Effect of substrates on formation and composition of poly-(R)-3-hydroxyalkanoates and poly-(R)-3-hydroxyalkenoates. *Appl. Environ. Microb.* **1988**, *54*, 2924–2932.
61. Scheiner, D. Determination of ammonia and Kjeldahl nitrogen by indophenol method. *Water Res.* **1976**, *10*, 31–36. [CrossRef]

62. Elbahloul, Y.; Steinbuhel, A. Large-scale production of poly(3-hydroxyoctanoic acid) by Pseudomonas putida GPo1 and a simplified downstream process. *Appl. Environ. Microb.* **2009**, *75*, 643–651. [CrossRef]
63. Kenny, S.T.; Runic, J.N.; Kaminsky, W.; Woods, T.; Babu, R.P.; Keely, C.M.; Blau, W.; O'Connor, K.E. Up-Cycling of PET (Polyethylene Terephthalate) to the Biodegradable Plastic PHA (Polyhydroxyalkanoate). *Environ. Sci. Technol.* **2008**, *42*, 7696–7701. [CrossRef]

*catalysts*

MDPI

*Article*

# Influence of Chemical Modifications of Polyhydroxyalkanoate-Derived Fatty Acids on Their Antimicrobial Properties

**Wojciech Snoch [1], Karolina Stępień [2], Justyna Prajsnar [1], Jakub Staroń [3], Maciej Szaleniec [1] and Maciej Guzik [1,*]**

[1] Jerzy Haber Institute of Catalysis and Surface Chemistry, Polish Academy of Sciences, Niezapominajek 8, 30-239 Kraków, Poland; ncsnoch@cyfronet.pl (W.S.); justyna.prajsnar1@gmail.com (J.P.); ncszalen@cyfronet.pl (M.S.)

[2] Medical University of Warsaw, Department of Pharmaceutical Microbiology, 3 Oczki Str. 02-007 Warsaw, Poland; Karolinastepien@windowslive.com

[3] Institute of Pharmacology, Polish Academy of Sciences, Smętna 12, 31-343 Kraków, Poland; jakubstaron@gmail.com

* Correspondence: ncguzik@cyfronet.pl; Tel.: +48-126-395-155

Received: 15 May 2019; Accepted: 30 May 2019; Published: 5 June 2019

check for updates

**Abstract:** Sugar esters are bioactive compounds derived from renewable resources. They consist of a sugar moiety with attached non-polar part – usually a fatty acid. These compounds find uses in cosmetic, food and pharmaceutical industries as surfactants due to their physicochemical and antimicrobial activities. In this study we have produced fatty acids for sugar ester synthesis from bacterially derived polyesters, namely polyhydroxyalkanoates (PHAs). We have developed methodology to decorate PHA monomers with a fluorinated moiety. With aid of biocatalysis a series of glucose esters was created with unmodified and modified PHA monomers. All synthesised compounds showed moderate antimicrobial activity.

**Keywords:** polyhydroxyalkanoate; (*R*)-3-hydroxyacids; biocatalysis; sugar esters; antimicrobial

## 1. Introduction

Sugar esters represent a class of biodegradable and bioactive compounds that brought attention of scientists and industries in recent decades [1]. These molecules are composed of a hydrophilic sugar component (rich in hydroxyl groups) connected via ester bonds to a hydrophobic part, usually a carboxylic acid. The number and the length of these acid derived chains, together with the number of hydroxyl groups of the particular sugar determine the hydrophilic-lipophilic balance (HLB) - a unique property of a given molecule of sugar ester. At the same time their chemical structures enable them to form micelles, emulsions, stabilize foams thanks to their surface active properties [2,3]. These features bring practical importance of sugar fatty acids esters (SFAEs), hence they are widely used in the food, cosmetic and pharmaceutical industries (as additives for dairy products, feeds, creams, gels, shampoos, pastes, ointments) [3–7]. SFAE not only have interesting physicochemical properties, but also are biologically active [8,9], increase the cell membrane permeability of microorganisms and inactivate transmembrane proteins, thus exposing microbes to adverse external factors, loss of intracellular proteins and valuable nutrients leading to an immediate inactivation [10,11].

The most popular among the sugar esters are those which contain glucose, galactose, sucrose or lactose because of their availability from natural resources [12–16]. The hydrophobic component is commonly derived from plant or animal biomass, e.g., butyric (C4), caprylic (C8), pelargonic (C9), lauric (C12), palmitic (C16) and stearic (C18) acids [5,8,9,17,18]. Depending on the length of aliphatic chain, its branching and existence of additional functional groups SFAE molecules may be more or less available for a given organism (due to variations in cell membrane transfer) and also faster or slower decomposed by enzymes (lipases, glycosidases). The presence of additional functional groups in the aliphatic chain modifies both the physical properties and therapeutic effect of SFAE, thus enhancing its biological effectiveness and spectrum [19–22]. These include phenyl, hydroxyl, amino, hydroxyphenyl, methoxy-phenyl and halogen-containing moieties, e.g., phenyl-fluorine, or alkyl halides [23–26]. Out of the mentioned modifiers, the fluorine most profoundly alters the molecule's biological functions (i.e., increases the free-radical production; influences bio-retention due to increased hydrophobicity of a given molecule) [27–29].

In order to aid in the search for new modifiers of the hydrophobic component of SFAE we tapped into monomeric units that build bacterially derived polymers. Bacteria synthesise polyhydroxyalkanoates (PHAs) in response to unfavourable environmental conditions employing sophisticated biochemical apparatus [30]. PHAs are composed of 3-hydroxylated fatty acids, where the hydroxyl group is always in an absolute *R* configuration [31]. Existence of this hydroxyl group on the carboxylic acid moiety opens interesting paths for modifications [32]. For example, it can be readily used to form an ether bond between oxygen atom and a halogenated alkyl group, which may influence their potential therapeutic effect. In recent years similar modifications of PHA monomers have been performed. However, they relied on the elimination of the hydroxyl group on chiral 3rd carbon atom and its exchange to 3-chloro, 3-fluorobenzyl, 3-bromo, 3-fluoro or amino group, which led to the loss of chirality of the molecule, but improved bacteriostatic properties [33,34].

Our study presents the development of a biocatalytic synthesis method for the preparation of unique glucose esters based on PHA derived monomers, namely mixtures of (*R*)-3-hydroxynonanoic and (*R*)-3-hydroxyheptanoic acids or (*R*)-3-hydroxy-5-phenylpentanoic and (*R*)-3-hydroxy-3-phenylpropionic acids arising from two types of PHA polymers obtained in bacterial fermentation process grown on, respectively, nonanoic or 5-phenylvaleric acids. The PHA derived acids were further functionalised with a 2,2,2-trifluoroethyl trifluoromethyl sulphate moiety. The virgin mixtures or their modified counterparts were attached to glucose via lipase mediated biocatalytic reaction in a water-free organic solvent systems. The resulting novel SFAEs were purified and characterised, and submitted to antimicrobial studies in order to elucidate their potential antimicrobial characteristics.

## 2. Results and Discussion

### 2.1. *Synthesis of Polyhydroxyalkanoates*

*P. putida* CA-3 was cultured on 5-phenylpentanoic acid (40 mM) in a shake flask for 5 days, which yielded 2.43 g/L of cell dry weight (CDW) mass with 57% (*w*/*w*) of PHA. The extracted PHA polymer (PHPV) composed of (*R*)-3-hydroxy-5-phenylpentanoic and (*R*)-3-hydroxy-3-phenylpropionic acids in a 94:6 molar ratio. In case of polyhydroxynonanoate (PHN) fermentation, *P. putida* KT244 strain was grown on nonanoic acid and accumulated 71% of the polymer in CDW, while the molar ratio of obtained (*R*)-3-hydroxynonanoic and (*R*)-3-hydroxyheptanoic acid monomers was 7:3.

### 2.2. *Modification of PHA Monomers*

The procedure described in Section 3 allowed us to obtain 1.26 g (48.5% conversion) of a pure fraction of desired products (**3**,**4**) – PHN derived mixture of hydroxyacids methyl esters. Structural analysis ($^1$H NMR, MS) of compounds revealed that 3-OH groups of PHN methyl esters were successfully protected (Figures S4 and S5 in Supplementary Material). Product **10** lost the methyl group during the ionization in MS, therefore it was identified as acid ([M-H]$^-$ 255 *m*/*z*) at retention time (Rt) of 2.3 min. An analogous situation occurred with the shorter protected PHA monomer (product **11**), (227 *m*/*z*, Rt = 1.9 min). Additionally, we observed a fragmentation ion (162.9 *m*/*z*), which, in accordance with the theoretical predictions of the fragmentation of compound **11** (Figure S5 in Supplementary Materials). The remaining chromatography fractions contained either dimers or trimers of **10**+**11** or their fluorinated counterparts. Unfortunately all trials for P3HPV derived fatty acid modifications ended without significant results. $^{14}$C NMR confirmed that an etheric bond between C–O$^-$ at the 3rd carbon atom of the P3HPV monomer aliphatic chain and trifluorethyl group was created. However purity of the collected fractions was unsatisfying, thus not allowing us from further biocatalytic steps with the mixture of **34+35**. Results of all were gathered and summarized in Table 1.

**Table 1.** Summary of PHA modifications and their sugar esters synthesis.

| Compound Number: | Conversion [%]: |
|---|---|
| **15** | 100 |
| **10,11** | 48.5 |
| **7,8** | 100 |
| **13,14** | n.c. |
| **26** | 42.1 * |
| **27,28** | 85 * |
| **29,30** | n.c. |
| **31** | 43.3 * |
| **32,33** | 78.7 * |
| **34,35** | n.c. |

* Conversions calculated from glucose concentrations.

To the best of our knowledge this is a first report that concerns protection of the hydroxyl group of a PHA monomers without its removal or its substitution by a halogen substituent [24,33–35]. NaH turned out to be an efficient reactant for hydrogen atoms removal of PHN 3-OH groups and their activation for an attack of trifluoroethyl moiety, even though lithium diisopropylamide is a much stronger nucleophil. Most importantly, an extra dry THF as a medium turned out to be the key element of the whole synthetic process.

Conversion of glucose to its corresponding esters (SFAE) via *Thermomyces lanuginosus* lipase (TL-IM) was confirmed by MS/MS analysis (Table 2). Compound **26** was identified as $[M+Cl^-]^-$ adduct 355 *m*/*z* in a scan mode, whereas in Product Ion Scan mode its precursor $[M-H]^-$ ion (319 *m*/*z*) produced fragmentation ion 228.9 *m*/*z* (Figure S6 in Supplementary Materials). PHN glucose esters (**27,28**) - both C9 and C7 were detected as $[M + Cl^-]^-$ ions in the scan mode, where their signals were 371 and 343 *m*/*z* respectively. In Product Ion Scan mode the precursor and fragmentation ions were observed at 335 and 307 *m*/*z*. Compound **27** in Product Ion Scan was characterized by fragmentation signals at 334.7, 172.8, 58.7 *m*/*z*. Such fragmentation was consistent with results of the theoretical analysis of the breakdown of the compound (Figures S7 and S8 in Supplementary Materials). Glucose esters of aromatic compounds have been synthesized with success, as indicated by their spectra in MS analysis. The product **31** was detected as the $[M-H]^-$ ion (339 *m*/*z*). In the case of PHPV sugar esters, only the MS spectrum with (R)-3-hydroxy-5-phenylpentanoic acid (compound **33**) was obtained. The compound was detected in the scan mode as the mass with the chlorine adduct, ($[M+Cl^-]^-$ = 391 *m*/*z*) with retention time of 1.4 min. In Product Ion Scan mode the **33** precursor ion $[M-H]^-$ (355 *m*/*z*) did not fragmented within the range of scanned energy collisions. The yield obtained within our study are comparable to these obtained for acylation of glucose by others [36,37].

### 2.3. Antimicrobial Testing

The initial microbial experiment showed that MIC (or Minimum bactericidal concentration—MBC) values of both unmodified and modified PHA monomers and their glucose esters are in most cases higher than 500 μg mL$^{-1}$, which is 50,000 fold higher than commonly used antibiotics (Ciplofloxan, Fluxonazole) [38,39]. Nevertheless, the literature reports and our results show that SFAEs in general have a bacteriostatic effect if present in higher concentrations, which can be used for other than therapeutic purposes (e.g., in the cosmetic or food industries, as surfactants/emulsifiers) [20,21,40,41]. Therefore, we increased the tested concentrations up to 5000 μg mL$^{-1}$ (Table 3).

**Table 2.** Precursor ions and Product ions of tested compounds in product ion scan mode.

| Compound | Precursor Ion (*m/z*) | Product Ions (*m/z*) | Collision Energies (eV) | Retention Time (min) |
|---|---|---|---|---|
| **1** | 157 | 157.1 | 5 | 1.6 |
| **3** | 173 | 173.1, 59 | 5 | 1.9 |
| **4** | 145 | 145.1, 59 | 5 | 1.8 |
| **5** | 191 | 114.8, 190.6 | 5 | 1.6 |
| **7** | 207 | 176.9, 206.9 | 5 | 1.6 |
| **8** | 179 | n.c | - | - |
| **12** | 255 | 255 | 5 | 2.3 |
| **13** | 227 | 226.7, 162.9 | 5 | 1.9 |
| **26** | 319 | 318.5, 228.9 | 5 | 1.7 |
| **27** | 335 | 334.7, 172.8, 58.7 | 10 | 1.4 |
| **28** | 307 | n.c. * | - | - |
| **29** | 417 | 417.1 | 25 | 1.6 |
| **30** | 389 | n.c | - | - |
| **31** | 339 | 338.9 | 5 | 1.7 |
| **33** | 356 | 354.9 | 5 | 1.4 |
| **32** | n.c. | n.d. * | - | - |

* n.c.—not confirmed, n.d.—not determined.

The results obtained during this study prove that microorganisms exhibit different sensitivity to antimicrobial agents, which strongly depends on the type of carbohydrate used, length of fatty acids attached to the sugar and moiety linked to the fatty acid part of SFAE. 5-phenylvaleric acid proved to be more effective against the tested bacteria at lower concentrations compared to aliphatic nonanoic acid. The antibacterial action of acids with the phenyl group has been known before—phenylpropionic or phenylacetic acid have a strong antibacterial effect on various types of bacteria such as *S. aureus*, *C. albicans* and *E. coli* at 1000 μg $mL^{-1}$ [42]. The opposite situation was observed in the case of inhibition of *Candida* growth, where nonanoic acid was more effective at 10 times lower concentrations compared to aromatic acids [43]. It has been reported that nonanoic acid could be active against some dermatophytes and non-dermatophyte fungi. Some of other saturated fatty acids (i.e., capric and lauric) have been also reported to possess the inhibitory effect against *Candida* spp. Fungicidal activity of fatty acids could be determined by mechanisms focused on yeast membrane disruption [43]. Sugar fatty acids esters also tend to be biologically beneficial in combating fungi. Their emulsifying properties could be successful implement in detergents and cosmetic production.

PHA monomers with a hydroxyl group at the 3rd carbon atom showed a stronger antibacterial activity (for bacteria strain such as *Staphylococcus aureus* NCTC 4163 and *Bacillus cereus* ATCC *11778)*, when compared to their fatty acids analogues without the 3-OH group. Similar results were obtained by Sandoval and colleagues, who tested the effect of the hydroxyl group on the antibacterial properties of PHA monomers [44]. For other strains tested no improvement of antibacterial activity was observed in the MIC between 5-phenylvaleric acid and PHPV derived monomer mixture. Attachment of the fluorinated moiety to PHN derived monomers did not significantly influenced their biological activity against tested strains.

**Table 3.** Activity of modified and unmodified PHA monomers and their SFAE against standard bacteria and fungi strains – minimal inhibitory concentrations (MIC, μg mL$^{-1}$).

| Ciprofloxacin/ Fluxonazole | [34,35] | [32,33] | [27,28] | [31] | [26] | [10,11] | [7,8] | [3,4] | [5] | [1] | Compound [ug mL$^{-1}$]: |
|---|---|---|---|---|---|---|---|---|---|---|---|
| 4 | 2500 | >5000 | 2500 | 5000 | 5000 | 2500 | 2500 | 1250 | 1250 | 2500 | *Staphylococcus aureus NCTC 4163* |
| 0.5 | 5000 | >5000 | 1250 | >5000 | 5000 | 2500 | 2500 | 2500 | 1250 | 2500 | *Staphylococcus aureus* ATCC 6538 |
| 0.5 | >5000 | >5000 | >5000 | >5000 | 5000 | 2500 | 2500 | 2500 | 1250 | 2500 | *Staphylococcus epidermidis* ATCC 12228 |
| 0.5 | 5000 | >5000 | 1250 | >5000 | 5000 | 2500 | 1250 | 2500 | 1250 | 2500 | *Staphylococcus epidermidis* RP 62A |
| 0.5 | >5000 | >5000 | >5000 | >5000 | 5000 | 2500 | 2500 | 2500 | 2500 | 2500 | *Enterococcus hirae ATCC 10541* |
| 0.5 | >5000 | >5000 | >5000 | >5000 | 2500 | 2500 | 2500 | 1250 | 1250 | 2500 | *Bacillus cereus* ATCC *11778* |
| 0.5 | >5000 | >5000 | >5000 | >5000 | 5000 | 2500 | 2500 | 2500 | 1250 | 2500 | *Bacillus subtilis* ATCC *6633* |
| 0.5 | >5000 | >5000 | >5000 | >5000 | >5000 | 2500 | 5000 | 5000 | 2500 | 5000 | *Escherichia coli* ATCC 25922 |
| 0.5 | >5000 | >5000 | >5000 | >5000 | >5000 | 5000 | 5000 | 5000 | 2500 | 5000 | *Pseudomonas aeruginosa* ATCC 27853 |
| 0.5 | >5000 | >5000 | >5000 | >5000 | 5000 | 2500 | 2500 | 5000 | 2500 | 2500 | *Salmonella enterica* subsp. *fnterica* CIP 108115 |
| 0.5 | >5000 | >5000 | 5000 | >5000 | 5000 | 2500 | 1250 | 2500 | 2500 | 2500 | *Listeria monocytogenes* |
| 0.5 | >5000 | >5000 | >5000 | >5000 | 313 | 5000 | >5000 | >5000 | 625 | 156 | *Candida parapsilosis* ATCC 22019 |
| 0.5 | >5000 | >5000 | >5000 | >5000 | 625 | 5000 | >5000 | 2500 | 1250 | 313 | *Candida albicans* ATCC 90028 |
| 0.5 | >5000 | >5000 | >5000 | >5000 | 625 | >5000 | >5000 | 5000 | 1250 | 156 | *Candida krusei* ATCC 6258 |
| 0.5 | >5000 | >5000 | >5000 | >5000 | 1250 | 5000 | >5000 | 5000 | 1250 | 156 | *Candida albicans* ATCC 10231 |

Sugar esters synthetized in this work revealed weak antibacterial to moderate antifungal activities (Table 3). C9-glucose ester (**26**) exhibited the highest antifungal activity of all studied sugar esters (MIC values of 313; 625; 625 μg $mL^{-1}$ for *C. parapsilosis; C. albicans ATCC 90028; C. crusei ATCC 1023*, respectively). We observed also some antibacterial activity of PHN glucose esters (**27,28**) towards *Staphylococcus* spp. in MIC range of 1250–2500 μg $mL^{-1}$. When modified with fluorine moiety, the PHN derived esters (**34,35**) retained their antimicrobial activity for *Staphylococcus aureus* NCTC 4163 (MIC 2500 μg $mL^{-1}$). For other tested pairs (glucose esters vs. tested strain) we did not observe significant antimicrobial action. These findings are in line with reports of others, where glucose or maltose sugar esters (fatty acid carbon atoms n = 8–14) revealed MIC values between 250 to 2000 μg $mL^{-1}$, when tested against a range of Gram negative and positive strains [10,45].

## 3. Materials and Methods

### 3.1. Synthesis of Polyhydroxyalkanoates

Polyhydroxynonanoate (PHN; **2**) was produced with *Pseudomonas putida* KT2440 strain with nonanoic acid (**1**) in the fermentation feed as the sole energy and carbon source (Figure 1A) as described in our previous study [46]. Polyhydroxyphenylvalerate (PHPV; **6**, Figure 1B) was obtained in shake flask cultures with *Pseudomonas putida* CA-3 (50 mL total medium volume in 250 mL Erlenmeyer flasks, 30 °C, 250 rpm, 5 day fed-batch cultivation) in Minimal Salt Medium (containing (g $L^{-1}$): $Na_2HPO_4{\cdot}12H_2O$, 9.0; $KH_2PO_4$, 1.5; $MgSO_4{\cdot}7H_2O$, 0.2; $NH_4Cl$, 1.0; $CaCl_2{\cdot}2H_2O$, 0.02; Fe(III)$NH_4$-citrate, 0.0012) with sodium phenylvalerate (total 80 mM). Polymers were extracted with ethyl acetate and characterised as described previously [46]. The polymers were degraded to their monomeric units via an acidic methanolysis (15% $H_2SO_4$ in methanol) to yield hydroxy fatty acid methyl esters according to previously established protocol [34].

### 3.2. Modification of PHA Monomers

Several methods of the 3-OH PHA methyl esters groups' halogenation were tested. Briefly different fluorinated alkyl iodes (1-fluoro-3-iodopropane; 1,1-difluoro-2-iodoethane; 1,1,1-trifluoro-3-iodopropane), various nucleophiles (sodium hydride (NaH; 60%), lithium diisopropylamide (LDA)) and different media (dimethylformamide (DMF), dichloromethane (DCM), tetrahydrofuran (THF)) were tested. Finally one method was chosen and was as follows. To stirred solution of 2.26 g PHN (**2**; 1 eq. mol) in anhydrous THF (20 mL), NaH (1.2 eq. mol) was added under argon atmosphere. After 30 min 2,2,2 -trifluoroethyl trifluoromethyl sulfate (**9**; 1.2 eq. mol) was added (Figure 1C). The reaction mixture was stirred on ice overnight. After acidification, an extraction in ethyl acetate (Et-Ac)/$H_2O$ was performed. TLC in Ac 1:7 H confirmed that no PHN (**2**) was left in the reaction mixture. Versa flash chromatography purification gave fraction of 1.5 g fluorinated PHN methyl esters (**10,11**) mixture ($^1H$ NMR spectra are available in supplementary materials). PHPV modification followed under the same procedure (Figure 1D). After modifications a part of all of the obtained monomers was converted into their acidic forms using *Candida antractica* lipase B (CalB) in water environment (Figure 1F,H). Briefly 300 mg of a given methyl ester was dissolved in 2 mL of dichloromethane. To 5 mL of $H_2O$, 85 mg of lipase was added. Both mixtures were combined and vigorously shaken (240 rpm, 35°C) overnight. Mixtures were acidified to pH 4 with HCl, and brine was added followed by the addition of 7 mL of ethyl acetate in order to extract (3×) the desired fatty acid.

### 3.3. SFAE Synthesis

Enzymatic reactions were carried out in water free 2-methyl-2-butanol (2M2B) in 10 mL total volume. Glucose was supplemented to yield 4 mg $mL^{-1}$ (1 eq. mol), whereas methyl nonaoate (**15**) to 11 mg $mL^{-1}$; PHN methyl esters (**3, 4**) 9.5 mg $mL^{-1}$; fluorinated PHN methyl esters (**10,11**) 11.46 mg $mL^{-1}$; 5-phenylpentanoic methyl esters (**20**) to 8.3 mg $mL^{-1}$; PHPV methyl esters (**7,8**) 9.02 mg (to 2 eq. mol). A quantity of 100 mg $mL^{-1}$ of molecular sieves (4Å) was added to withdraw the water formed. Reactions were initiated by addition of 12 mg $mL^{-1}$ of an enzyme (*Thermomyces lanuginosus* lipase herein referred to as TL-IM), with shaking (240 rpm), at 55 °C for 24 h (New Brunswick Scientific Exella E24 Incubator Shaker Series; (Figure 2)) [36,37,47]. Samples were collected at set intervals and were analysed on HPLC-MS system.

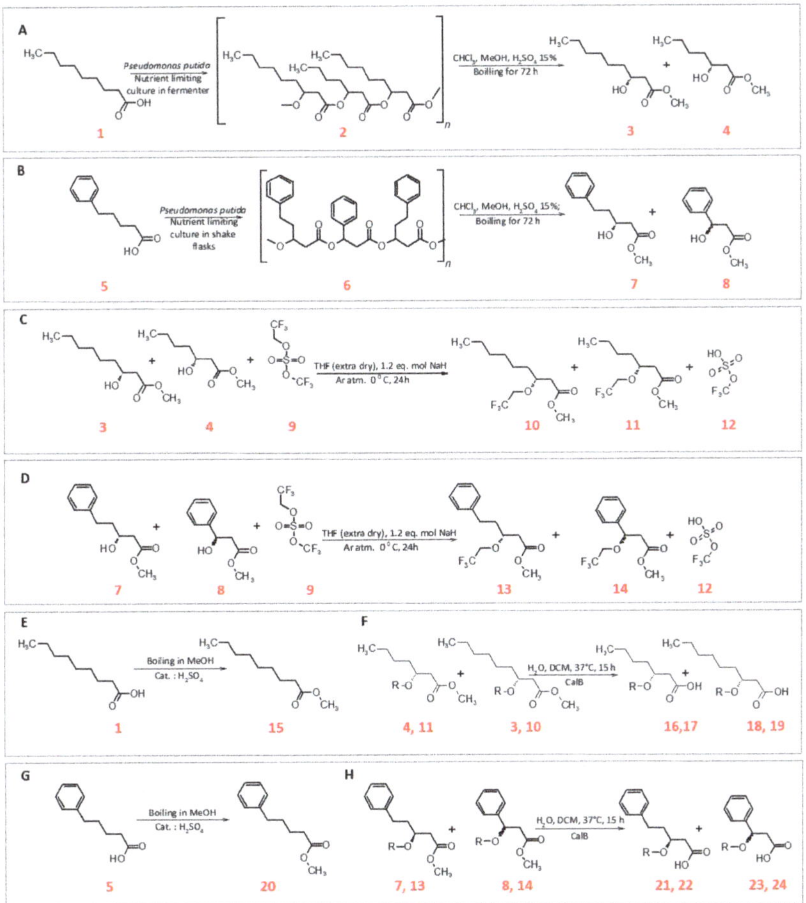

**Figure 1. Scheme of synthesis PHA and modifications their monomers** (**A+B**: PHN synthesis and methanolysis; **C+D**: Fluorination of PHN monomers **E+G**: methyl ester synthesis; **F+H**: demethylation of substrates with CalB); **1.** nonanoic acid; **2.** poly-3-hydroxyhydroxynonaoate; **3.** methyl (3R)-3-hydroxynonanoate; **4.** methyl (3R)-3-hydroxyheptanoate; **5.** 5-phenylpentanoic acid; **6.** poly-3-hydroxypentanoate; **7.** methyl (3R)-3-hydroxy-5-phenylpentanoate; **8.** methyl (3R)-3-hydroxy-3-phenylpropanoate; **9.** 2,2,2-trifluoroethyl trifluoromethyl sulfate; **10.** methyl (3R)-3-(2,2,2-trifluoroethoxy)nonanoate; **11.** methyl (3R)-3-(2,2,2-trifluoroethoxy)heptanoate; **12.** trifluoromethyl hydrogen sulfate; **13.** methyl (3R)-5-phenyl-3-(2,2,2-trifluoroethoxy)pentanoate; **14.** methyl 3-phenyl-3-(2,2,2-trifluoroethoxy)propanoate; **15.** methyl nonanoate; **16.** (R)-3-hydroxyheptanoic acid; **17.** 3-(2,2,2-trifluoroethoxy)heptanoic acid; **18.** (R)-3-hydroxynonanoic acid; **19.** 3-(2,2,2-trifluoroethoxy)nonanoic acid; **20.** methyl 5-phenylpentanoate; **21.** (R)-3-hydroxy-5-phenylpentanoic acid; **22.** 5-phenyl-3-(2,2,2-trifluoroethoxy)pentanoic acid; **23.** (R)-3-hydroxy-3-phenylpropanoic acid; **24.** 3-phenyl-3-(2,2,2-trifluoroethoxy)propanoic acid.

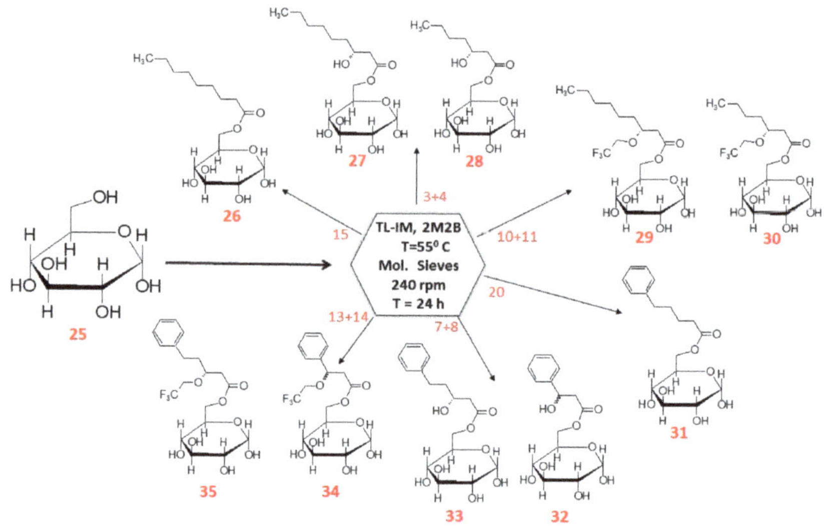

**Figure 2. Scheme of synthesis SFAE. 25.** Glucose [(2*S*,5*S*)-6-(hydroxymethyl) oxane-2,3,4,5-tetrol]; **26.** [(3*S*,6*S*)-3,4,5,6- tetrahydroxyoxan-2-yl] methyl nonanoate]; **27.** [(3*S*,6*S*)-3,4,5,6-tetrahydroxyoxan-2-yl] methyl (3*R*)-3-hydroxynanonoate; **28.** [(3*S*,6*S*)-3,4,5,6-tetrahydroxyoxan-2-yl] methyl (3*R*) -3-hydroxyheptanoate; **29.** [(3*S*,6*S*)-3,4,5,6-tetrahydroxyoxan-2-yl] methyl(3*R*)-3-(2,2,2-trifluoroethoxy) nonanoate; **30.** [(3*S*,6*S*)-3,4,5,6-tetrahydroxyoxan-2-yl] methyl (3*R*)-3-(2,2,2-trifluoroethoxy) heptanoate; **31.** [(3*S*,6*S*)-3,4,5,6-tetrahydroxyoxan-2-yl] methyl 5-phenylpentanoate; **32.** [(3*S*,6*S*)-3,4,5,6-tetrahydroxyoxan-2-yl]methyl (3*R*) -3-hydroxy-3-phenylpropanoate; **33.** [(3*S*,6*S*)-3,4,5,6-tetrahydroxyoxan-2-yl] methyl (3*R*)-3-hydroxy-5-phenylpentanoate; **34.** [(3*S*,6*S*)-3,4,5,6-tetrahydroxyoxan-2-yl] methyl (3*R*)-3-phenyl-3-(2,2,2-trifluoroethoxy) propanoate; **35.** [(3*S*,6*S*)-3,4,5,6-tetrahydroxyoxan-2-yl] methyl (3*R*) -3-phenyl-3-(2,2,2-trifluoroethoxy) pentanoate.

### *3.4. LC-MS*

The analyses were performed by UHPLC measurements on Agilent 1290 Infinity System with automatic autosampler and MS Agilent 6460 Triple Quad Detector (Santa Clara, CA, USA) equipped with Agilent Zorbax Eclipse Plus C18 column (2.1 × 50 mm, 1.8 μm). Separations were conducted at 30 °C at in a gradient of water (A) and methanol (B) according to the eluent program 0.00 min (95% A/5% B) to 1.00 min (100% B) to 3.51 min (95% A/5% B) to 4.50 min (95% A/5% B) using a flow rate of 0.4 mL $min^{-1}$. The 5 μl injections of samples were applied in duplicates. An MS Agilent 6460 Triple Quad tandem mass spectrometer with an Agilent Jet Stream ESI interface was used in negative ion polarization using either the Scan (MS2 scan) or Product Ion Scan modes. The optimum collision energy was 5 eV for all of the products. Nitrogen at a flow rate of 10 L $min^{-1}$ was used as the drying gas and for collision-activated dissociation. The drying gas and sheath gas temperatures were set to 350 °C. The capillary voltage was set to 3500 V, whereas the nozzle voltage was set to 500 V. All compounds were monitored in scan and product ion modes with different collision energies (5–30 eV) (*m*/*z* values of products in the Table 2). MassHunter software (Version B.05.00, Agilent, Santa Clara, CA, USA) was used for HPLC-MS system control, data acquisition, and data processing.

### *3.5. Antimicrobial Testing*

The studies of antimicrobial activity were conducted for modified and non-modified PHA monomers (acids) and their SFAE derivatives using clinical and reference strains of bacteria and yeast-like fungi from international microbe collections: ATCC (American Type Culture Collection), NCTC (National Collection of Type Culture) and CIP (The Collection de l'Institut Pasteur). Among reference strains, there were seven Gram-positive bacteria (*Staphylococcus aureus* NCTC 4163, *S. aureus* ATCC 6538, *S. epidermidis* ATCC 12228, *S. epidermidis* ATCC 35984, *Enterococcus hirae* ATCC 10541, *Bacillus cereus* ATCC 17778 and *B. subtilis* ATCC 6633) and four Gram-negative bacteria (*Pseudomonas aeruginosa* ATCC 27853, *Escherichia coli* ATCC 25922, *Salmonella enterica* subsp. *enterica* CIP 108115 and clinical isolate of *Listeria monocytogenes*). The yeast-like fungi used in this study were *Candida* spp. (*C. parapsilosis* ATCC 22019, *C. krusei* ATCC 6258, *C. albicans* ATCC 90028 and *C. albicans* ATCC 10231). Compounds antimicrobial activity was expressed as minimum inhibitory concentration (MIC) according to The European Committee on Antimicrobial Susceptibility Testing (EUCAST) and The Clinical & Laboratory Standards Institute (CLSI) reference procedures with some modification. MIC was tested by the twofold serial microdilution method (in 96-well microtiter plates) on MH II liquid medium for bacteria or RPMI- 1640 medium for *Candida* species. The final inoculum of all studies bacteria was $10^{6\circ}$ CFU/mL (colony forming unit per mL) and $5 \times 10^4$ to $2.5 \times 10^5$ CFU $mL^{-1}$ for yeast. The stock solutions of tested compounds were prepared in DMSO and diluted in sterile medium (to maximum 3% of solvent content). The concentrations of compounds were from 78 to 5000 $\mu g\ mL^{-1}$. The MIC value was the lowest concentration of the researched compound at which bacteria growth was no longer observed after 18 h. Yeast growth was evaluated by absorbance measurement at 530 nm after at least 24 h of incubation. The MIC was defined as a 50% or more reduction in growth compared to the control well [38,48]. As controls, two antimicrobial compounds were used: ciprofloxacin (antibacterial) and fluxonazole (antifungal).

## 4. Conclusions

This work provides an insight into synthesis of novel range of compounds derived from bacterial polyesters – polyhydroxyalkanoates. Firstly, we have established a methodology for protection of hydroxyl group of PHA derived monomers by introducing a fluorinated moiety via an etheric bond. Secondly, we developed a protocol for biocatalytic acylation of glucose with these bacterially derived fatty acids. Further in the study we have tested antimicrobial potential of these unmodified and modified PHA derivatives. The obtained compounds revealed moderate antibacterial and antifungal activities. Further research is needed in order to increase the antimicrobial activity of either PHA monomers or their sugar esters by introduction of other bioactive structural components to their moieties.

**Supplementary Materials:** The following are available online at http://www.mdpi.com/2073-4344/9/6/510/s1.

**Author Contributions:** Conceptualization, M.G.; Data curation, J.S.; Formal analysis, W.S.; Investigation, W.S., K.S. and J.P.; Supervision, J.S., M.S. and M.G.; Visualization, J.P.; Writing—original draft, W.S., K.S. and J.P.; Writing—review & editing, J.S., M.S. and M.G.

**Funding:** This work was supported by the National Science Centre, Poland [grant SONATA no. 2015/17/D/ST4/00514]. JP and WS acknowledge the support of InterDokMed project no. POWR.03.02.00-00-I013/16. We greatly acknowledge the joint consortium "Interdisciplinary Centre of Physical, Chemical and Biological Sciences" of ICSC PAS and INP PAS for providing access to the Agilent 1290 Infinity System with an automatic autosampler and an MS Agilent 6460 Triple Quad Detector. The research on antimicrobial activity was carried out with the use of CePT infrastructure financed by the European Union—the European Regional Development Fund within the Operational Programme "Innovative economy" for 2007–2013.

**Conflicts of Interest:** The authors declare no conflicts of interest.

## References

1. Staroń, J.; Dąbrowski, J.M.; Cichoń, E.; Guzik, M. Lactose esters: Synthesis and biotechnological applications. *Crit. Rev. Biotechnol.* **2018**, *38*, 245–258. [CrossRef] [PubMed]
2. Hidayat, C.; Fitria, K.; Hastuti, P. Enzymatic synthesis of bio-surfactant fructose oleic ester using immobilized lipase on modified hydrophobic matrix in fluidized bed reactor. *Agric. Agric. Sci. Procedia* **2016**, *9*, 353–362. [CrossRef]

3. Jeromin, G.E.; Zoor, A.; Stergiou, P.-Y.; Foukis, A.; Filippou, M.; Koukouritaki, M.; Parapouli, M.; Theodorou, L.G.; Hatziloukas, E.; Afendra, A.; et al. Enzymatic esterification of tapioca maltodextrin fatty acid ester. *Carbohydr. Polym.* **2001**, *99*, 2079–2090.
4. Maag, H. Fatty acid derivatives: Important surfactants for household, cosmetic and industrial purposes. *J. Am. Oil Chem. Soc.* **1984**, *61*, 259–267. [CrossRef]
5. Hill, K.; Rhode, O. Sugar-based surfactants for consumer products and technical applications. *Lipid/Fett* **1999**, *101*, 25–33. [CrossRef]
6. Neta, N.D.A.S.; Santos, J.C.S.D.; Sancho, S.D.O.; Rodrigues, S.; Gonçalves, L.R.B.; Rodrigues, L.R.; Teixeira, J.A. Enzymatic synthesis of sugar esters and their potential as surface-active stabilizers of coconut milk emulsions. *Food Hydrocoll.* **2012**, *27*, 324–331. [CrossRef]
7. Zhao, T.-H.; Gu, J.-Y.; Pu, W.-F.; Dong, Z.-M.; Liu, R. Study on the synthesis and properties of an eco-friendly sugar-based anionic–nonionic surfactant. *RSC Adv.* **2016**, *6*, 70165–70173. [CrossRef]
8. El-Laithy, H.M.; Shoukry, O.; Mahran, L.G. Novel sugar esters proniosomes for transdermal delivery of vinpocetine: Preclinical and clinical studies. *Eur. J. Pharm. Biopharm.* **2011**, *77*, 43–55. [CrossRef]
9. Szuts, A.; Szabó-Révész, P. Sucrose esters as natural surfactants in drug delivery systems—A mini-review. *Int. J. Pharm.* **2012**, *433*, 1–9. [CrossRef]
10. Zhao, L.; Zhang, H.; Hao, T.; Li, S. In vitro antibacterial activities and mechanism of sugar fatty acid esters against five food-related bacteria. *Food Chem.* **2015**, *187*, 370–377. [CrossRef]
11. Xiao, D.; Ye, R.; Davidson, P.M.; Hayes, D.G.; Golden, D.A.; Zhong, Q. Sucrose monolaurate improves the efficacy of sodium hypochlorite against escherichia coli O157: H7 on spinach. *Int. J. Food Microbiol.* **2011**, *145*, 64–68. [CrossRef] [PubMed]
12. Ferla, B.L.; Lay, L.; Poletti, L.; Russo, G.; Panza, L. Easy chemo-enzymatic synthesis of human milk trisaccharides from a common selectively protected lactose building block. *J. Carbohydr. Chem.* **2000**, *19*, 331–343. [CrossRef]
13. Rencurosi, A.; Poletti, L.; Panza, L.; Lay, L. Improvement on lipase catalysed regioselective O-acylation of lactose: A convenient route to 2-O-fucosyllactose. *J. Carbohydr. Chem.* **2001**, *20*, 761–765. [CrossRef]
14. Desbois, A.P. Potential applications of antimicrobial fatty acids in medicine, agriculture and other industries. *Recent Pat. Antiinfect. Drug Discov.* **2012**, *7*, 111–122. [CrossRef] [PubMed]
15. Karlova, T.; Poláková, L.; Šmidrkal, J.; Filip, V. Antimicrobial effects of fatty acid fructose esters. *Czech J. Food Sci.* **2010**, *28*, 146–149. [CrossRef]
16. Das, B.; Sarkar, S.; Sarkar, A.; Bhattacharjee, S.; Bhattacharjee, C. Recovery of whey proteins and lactose from dairy waste: A step towards green waste management. *Process Saf. Environ. Prot.* **2016**, *101*, 27–33. [CrossRef]
17. Huang, C.B.; George, B.; Ebersole, J.L. Antimicrobial activity of n-6, n-7 and n-9 fatty acids and their esters for oral microorganisms. *Arch. Oral Biol.* **2010**, *55*, 555–560. [CrossRef]
18. Blanchfield, J.; Toth, I. Lipid, Sugar and Liposaccharide based delivery systems 2. *Curr. Med. Chem.* **2012**, *11*, 2375–2382. [CrossRef]
19. Zheng, C.J.; Yoo, J.S.; Lee, T.G.; Cho, H.Y.; Kim, Y.H.; Kim, W.G. Fatty acid synthesis is a target for antibacterial activity of unsaturated fatty acids. *FEBS Lett.* **2005**, *579*, 5157–5162. [CrossRef]
20. Smith, A.; Nobmann, P.; Henehan, G.; Bourke, P.; Dunne, J. Synthesis and antimicrobial evaluation of carbohydrate and polyhydroxylated non-carbohydrate fatty acid ester and ether derivatives. *Carbohydr. Res.* **2008**, *343*, 2557–2566. [CrossRef]
21. Watanabe, T.; Katayama, S.; Matsubara, M.; Honda, Y.; Kuwahara, M. Antibacterial carbohydrate monoesters suppressing cell growth of streptococcus mutans in the presence of sucrose. *Curr. Microbiol.* **2000**, *41*, 210–213. [CrossRef] [PubMed]
22. Galbraith, H.; Miller, T.B. Effect of metal cations and PH on the antibacterial activity and uptake of long chain fatty acids. *J. Appl. Bacteriol.* **1973**, *36*, 635–646. [CrossRef] [PubMed]
23. Fernandez-Lorente, G.; Palomo, J.M.; Cocca, J.; Mateo, C.; Moro, P.; Terreni, M.; Fernandez-Lafuente, R.; Guisan, J.M. Regio-selective deprotection of peracetylated sugars via lipase hydrolysis. *Tetrahedron* **2003**, *59*, 5705–5711. [CrossRef]
24. Dembitsky, V.M.; Srebnik, M. Natural halogenated fatty acids: Their analogues and derivatives. *Prog. Lipid Res.* **2002**, *41*, 315–367. [CrossRef]

25. Andreu, C.; Marcel, Æ.; Varea, T.; Diaz, D.; Asensio, G. The introduction of fluorine atoms or trifluoromethyl groups in short cationic peptides enhances their antimicrobial activity. *Bioorgan. Med. Chem.* **2006**, *14*, 6971–6978.
26. Hiyama, T.; Yamamoto, H. Biologically active organofluorine compounds. In *Organofluorine Compounds*; Springer: Berlin/Heidelberg, Germany, 2012; pp. 137–182.
27. Isanbor, C.; Hagan, D.O. Fluorine in medicinal chemistry: A review of anti-cancer agents §. *J. Fluor. Chem.* **2006**, *127*, 303–319. [CrossRef]
28. Cheeseman, K.H.; Albano, E.F.; Tomasi, A.; Slater, T.F. Biochemical studies on the metabolic activation of halogenated alkanes. *Environ. Health Perspect.* **1985**, *64*, 85–101. [CrossRef]
29. Belli, W.A.; Buckley, D.H.; Marquis, R.E. Weak acid effects and fluoride inhibition of glycolysis by streptococcus mutans GS-5. *Can. J. Microbiol.* **1992**, *41*, 785–791. [CrossRef]
30. Suriyamongkol, P.; Weselake, R.; Narine, S.; Moloney, M.; Shah, S. Biotechnological approaches for the production of polyhydroxyalkanoates in microorganisms and plants—A review. *Biotechnol. Adv.* **2007**, *25*, 148–175. [CrossRef]
31. Abe, H.; Doi, Y. Side-Chain effect of second monomer units on crystalline morphology thermal properties and enzymatic degradability for random copolyesters of (R) -3-hydroxybutyric acid with (R) -3-hydroxyalkanoic acids. *Biomacromolecules* **2002**, *3*, 133–138. [CrossRef]
32. Madison, L.; Huisman, G. Metabolic engineering of poly(3-hydroxyalkanoates): From DNA to plastic. *Microbiol. Mol. Biol. Rev.* **1999**, *63*, 21–53. [PubMed]
33. O'Connor, S.; Szwej, E.; Nikodinovic-Runic, J.; O'Connor, A.; Byrne, A.T.; Devocelle, M.; O'Donovan, N.; Gallagher, W.M.; Babu, R.; Kenny, S.T.; et al. The anti-cancer activity of a cationic anti-microbial peptide derived from monomers of polyhydroxyalkanoate. *Biomaterials* **2013**, *34*, 2710–2718. [CrossRef] [PubMed]
34. Radivojevic, J.; Skaro, S.; Senerovic, L.; Vasiljevic, B.; Guzik, M.; Kenny, S.T.; Maslak, V.; Nikodinovic-Runic, J.; O'Connor, K.E.; O'Connor, K. Polyhydroxyalkanoate-based 3-hydroxyoctanoic acid and its derivatives as a platform of bioactive compounds. *Appl. Microbiol. Biotechnol.* **2015**, *100*, 161–172. [CrossRef] [PubMed]
35. Constantin, M.; Simionescu, C.I.; Carpov, A.; Samain, E.; Driguez, H. Chemical modification of poly (hydroxyalkanoates). Copolymers bearing pendant sugars. *Macromol. Rapid Commun.* **1999**, *94*, 91–94. [CrossRef]
36. Walsh, M.K.; Bombyk, R.A.; Wagh, A.; Bingham, A.; Berreau, L.M. Synthesis of lactose monolaurate as influenced by various lipases and solvents. *J. Mol. Catal. B Enzym.* **2009**, *60*, 171–177. [CrossRef]
37. Plou, F.J.; Cruces, M.A.; Ferrer, M.; Fuentes, G.; Pastor, E.; Bernabé, M.; Christensen, M.; Comelles, F.; Parra, J.L.; Ballesteros, A. Enzymatic acylation of di- and trisaccharides with fatty acids: Choosing the appropriate enzyme, support and solvent. *J. Biotechnol.* **2002**, *96*, 55–66. [CrossRef]
38. Clinical and Laboratory Standards Institute (CLSI); Weinstein, M.P. *Methods for Dilution Antimicrobial Susceptibility Tests for Bacteria That Grow Aerobically*, 9th ed.; Clinical and Laboratory Standards Institute: Wayne, NY, USA, 2012.
39. Kümmerer, K. Antibiotics in the aquatic environment—A. review—Part I. *Chemosphere* **2009**, *75*, 417–434. [CrossRef]
40. Wagh, A.; Walsh, M.K.; Martini, S. Effect of lactose monolaurate and high intensity ultrasound on crystallization behavior of anhydrous milk fat. *J. Am. Oil Chem. Soc.* **2012**, *90*, 977–987. [CrossRef]
41. Lucarini, S.; Fagioli, L.; Campana, R.; Cole, H.; Duranti, A.; Baffone, W.; Vllasaliu, D.; Casettari, L. Unsaturated fatty acids lactose esters: cytotoxicity, permeability enhancement and antimicrobial activity. *Eur. J. Pharm. Biopharm.* **2016**, *107*, 88–96. [CrossRef]
42. Bills, G.; Cueva, C.; Moreno-arribas, M.V.; Martı, P.J.; Vicente, M.F.; Basilio, A.; Rodrı, J.M. Antimicrobial activity of phenolic acids against commensal, probiotic and pathogenic bacteria. *Res. Microbiol.* **2010**, *161*, 372–382.
43. Pohl, C.H.; Kock, J.L.F.; Thibane, V.S. Antifungal free fatty acids: A review antifungal free fatty acids: A review. *Sci. Microb. Pathog. Curr. Res. Technol. Adv.* **2011**, *1*, 61–71.
44. Sandoval, Á.; Arias-Barrau, E.; Bermejo, F.; Cañedo, L.; Naharro, G.; Olivera, E.R.; Luengo, J.M. Production of 3-hydroxy-n-phenylalkanoic acids by a genetically engineered strain of Pseudomonas putida. *Appl. Microbiol. Biotechnol.* **2005**, *67*, 97–105. [CrossRef] [PubMed]

45. Ferrer, M.; Soliveri, J.; Plou, F.J.; López-Cortés, N.; Reyes-Duarte, D.; Christensen, M.; Copa-Patiño, J.L.; Ballesteros, A. Synthesis of sugar esters in solvent mixtures by lipases from Thermomyces lanuginosus and Candida antarctica B, and their antimicrobial properties. *Enzyme Microb. Technol.* **2005**, *36*, 391–398. [CrossRef]
46. Sofińska, K.; Barbasz, J.; Witko, T.; Dryzek, J.; Haraźna, K.; Witko, M.; Kryściak-Czerwenka, J.; Guzik, M. Structural, Topographical, and Mechanical characteristics of purified polyhydroxyoctanoate polymer. *J. Appl. Polym. Sci.* **2019**, *136*, 47192. [CrossRef]
47. Siebenhaller, S.; Gentes, J.; Infantes, A.; Muhle-Goll, C.; Kirschhöfer, F.; Brenner-Weiß, G.; Ochsenreither, K.; Syldatk, C. Lipase-Catalyzed Synthesis of Sugar Esters in Honey and Agave Syrup. *Front. Chem.* **2018**, *6*, 1–9. [CrossRef]
48. Arendrup, M.C.; Meletiadis, J.; Mouton, J.W.; Lagrou, K.; Hamal, P.; Guinea, J. Subcommittee on Antifungal Susceptibility Testing of the ESCMID European Committee for Antimicrobial Susceptibility Testing. 2017 EUCAST Definitive Document E.DEF 9.3.1: Method for the Determination of Broth Dilution Minimum Inhibitory Concentrations of Antifungal Agents for Conidia Forming Moulds. Available online: http://www.eucast.org/fileadmin/src/media/PDFs/EUCAST_files/AFST/Files/EUCAST_E_Def_9_3_1_Mould_testing__definitive.pdf (accessed on 25 April 2019).

catalysts

MDPI

Article

# Nitrogen Removal by Co-Immobilized Anammox and Ammonia-Oxidizing Bacteria in Wastewater Treatment

Igor Dolejš [1], Radek Stloukal [2], Michal Rosenberg [1] and Martin Rebroš [1,*]

[1] Institute of Biotechnology, Faculty of Chemical and Food Technology, Slovak University of Technology in Bratislava, Radlinského 9, 812 37 Bratislava, Slovakia; igor.dolejs@gmail.com (I.D.); michal.rosenberg@stuba.sk (M.R.)

[2] LentiKat´s a.s., Pod Vinicí 83, 471 27 Stráž pod Ralskem, Czech Republic; radek.stloukal@seznam.cz

* Correspondence: martin.rebros@stuba.sk; Tel.: +421-2-59-325-480

Received: 4 May 2019; Accepted: 7 June 2019; Published: 12 June 2019

**Abstract:** In wastewater treatment, an alternative to the widely used aerobic nitrification with subsequent anoxic denitrification method is the combination of nitration and anammox (AMX) in one system. This study focuses on the co-immobilization of AMX and ammonia-oxidizing bacteria into a polyvinyl alcohol (PVA) hydrogel, and its effective use in nitrogen removal (NR). The NR process was performed in nine consecutive, repeated batches. By optimizing the conditions of the biotransformations, there was equal utilization of nitrogen in both sources, N–$NH_4^+$ and N–$NO_2^-$, at 100% NR during the sixth repetition. A significant increase in the immobilized co-culture activity was also detected per cycle. The maximum value of the NR rate was 3.46 mg N $(L\ h)^{-1}$, and 100% NR efficiency was achieved with an initial concentration of 100.3 mg N $L^{-1}$ for N–$NH_4^+$ and 60.1 mg N $L^{-1}$ for N–$NO_2^-$, during the eighth batch biotransformation.

**Keywords:** anammox; immobilization; wastewater treatment; polyvinyl alcohol

## 1. Introduction

Organic nitrogen-containing compounds are among the most important pollutants of wastewater, and the removal of nitrogen is one of the crucial steps in wastewater treatment. These compounds represent a main factor in eutrophication, and they have an effect on the oxygen content of receiving waters as well as the toxicity to aquatic organisms and human beings [1,2]. For decades, a removal process based on aerobic nitrification with subsequent anoxic denitrification was used for the removal of such nitrogen-based compounds. The discovery of anaerobic oxidation of ammonium (anammox; AMX), with essential advantages of high nitrogen removal rate (NRR), environmental friendliness, low operational costs and low occupied areas, has been recognized as an attractive alternative for the treatment of nitrogen-rich wastewater streams [3–5]. The AMX process operates by oxidizing ammonia to nitrogen gas with nitrite as an electron acceptor under anoxic conditions. Meanwhile, the growth of AMX bacteria is supported by carbon dioxide fixation (Equation (1)), which is an advantage for the treatment of highly nitrogen-loaded wastewaters containing low biodegradable organic carbons [6–8].

$$NH_4^+ + 1.3NO_2^- + 0.066HCO_3^- + 0.13H^+ \rightarrow 1.02\ N_2 + 0.26NO_3^- + 0.066CH_2O_{2.5}N_{0.15} + 2.03\ H_2O \quad (1)$$

As nitrite is needed for the AMX process at a molar ratio of 1:1.32 with ammonium [9], part of the ammonium in wastewater must be oxidized to nitrite with a pre-treatment system, such as the nitrosation process, realized by ammonia oxidizing bacteria (AOB), as shown in Equation (2) [6,7,10].

$$2NH_4^+ + 3O_2 \rightarrow 2NO_2^- + 2\,H_2O + 4H^+ \quad (2)$$

Nevertheless, compared to conventional nitrification and denitrification, AMX consumes 100% less organic carbon and saves 90% of the operational costs associated with sludge disposal [11]. An alternative pre-treatment system for wastewater is the combination of nitrosation and AMX in one system. So far, it has mainly been applied for wastewaters with high concentrations of ammonia and low concentrations of biodegradable organic substances [12]. Functional AMX bacteria are very sensitive and are easily inhibited by many factors, such as low temperature (optimal range is 20–43 °C), extreme pH value (optimum pH 6.5–8.8), high salinity as well as the presence of organic matter, phosphates, sulfides and other inhibitors [4,7].

For this reason, immobilization of microbial cells has received increasing interest in wastewater treatment in order to minimize the risk of biomass wash-out from the reactors and to provide a stabilized treatment [13]. The repeated use of immobilized biomass could handle the long start-up of AMX bacteria because of the very slow growth rate (0.072/day at 32 °C) and low yield coefficient (0.13 g dry weight/g $NH_4$–N oxidized) [11]. Immobilization of AMX bacteria through the entrapment method has been reported in several previous papers, using polyethylene glycol gel carriers [14,15] and polyvinyl alcohol (PVA) cryogel, prepared by physical cross-linking through the freezing/thawing method [13].

The aim of this research was to evaluate the effectiveness of co-immobilization of AMX bacteria and AOB by the entrapment method, immobilizing both bacteria into a PVA hydrogel, and its application in nitrogen removal (NR) during the water treatment process, using a real medium. The key issue in this application, however, is the harmonization of activities of both cultures. Therefore, we have focused on different process aspects to achieve this in co-immobilized form. Compared to other gel systems, immobilization using this method offers several advantages, such as a low matrix cost, inexpensive and simple gel preparation, uncomplicated separation from the reaction mixture and low diffusion limits. In addition, this matrix has excellent mechanical stability and is almost non-degradable [16].

## 2. Results and Discussion

The NR process using immobilized microbial co-cultures of AMX bacteria and AOB occurred in repeated batch-mode biotransformations. This experimental setup was made mainly because of the separate influences on each batch process (such as pH maintenance, nutrient limitation) which were evaluated after each separate batch. Another inspiration was to reach high NR efficiency. There were nine successive repetitions realized, as illustrated in Figure 1. The operational parameters were improved throughout the entire process and resulted in an increase of the nitrogen removal rate (NRR) and achieved a 100%-successful transformation of substrate nitrogen compounds contained in the medium.

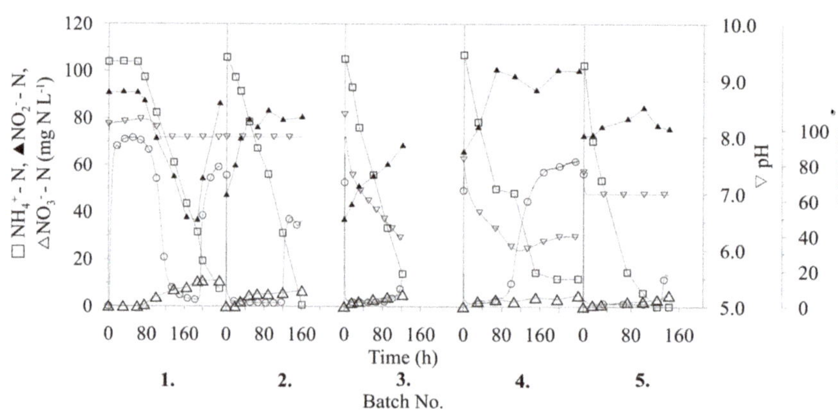


**Figure 1.** *Cont.*

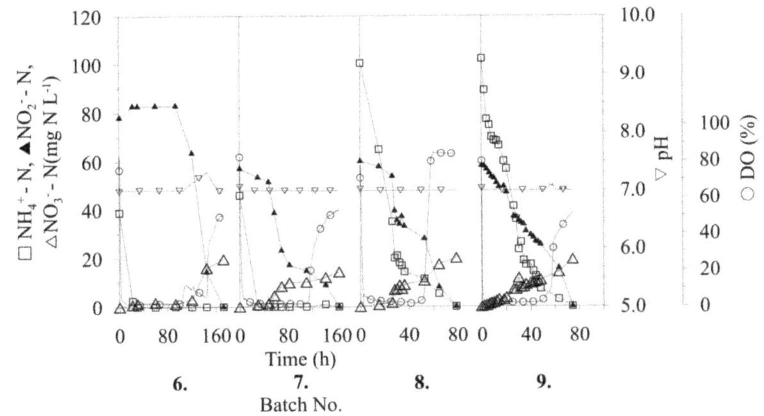


**Figure 1.** Repeated batch biotransformations with immobilized co-culture of anammox (AMX) bacteria and ammonia oxidizing bacteria AOB.

Figure 1 illustrates the process of each cycle of repeated biotransformation, and, as shown, low activity of the immobilized cells was observed during the first 60 h of the first biotransformation. This was most likely caused by the highly stressful aerobic conditions of the immobilization process to which the biomass was exposed for a short time beforehand. After this time, the oxygen started to decrease in the medium, as observed by a decrease in the dissolved oxygen (DO) content to 10%, presumably caused by production of nitrogen gas (a final compound of nitrogen conversion). The higher activity of the immobilized bacteria also presented simultaneous utilization of N–$NH_4^+$ and N–$NO_2^-$, probably caused by the suppression of the metabolic activity of one or more representatives of the immobilized AMX consortium also presented by formation of ions N–$NO_3^-$. Utilization of N–$NO_2^-$ slowed around the middle of the first repeated batch and, instead of decreasing, it started to increase, according to the fermentation time. An increase of N–$NO_2^-$ concentration and continuous utilization of N–$NH_4^+$ was more likely caused by the activity of AOB, which became dominant in the co-immobilized culture. As reported [17], AOB forms a thick layer around anammox cells which, after AOB domination, may eliminate the substrate (ammonia) access to AMX consortium (Equations (1,2)). During this first batch biotransformation, 52% of the total nitrogen was removed from compounds. The NRR, calculated from 0 to 30 h for each batch, was constant at 0 mg N $(L\ h)^{-1}$ because of the long lag phase. After 256 h of the first biotransformation, the whole volume of the medium was separated through a sieve, subsequently fed with fresh medium and the next biotransformation was started.

As shown in Figure 1, the trend of nitrite production continued during the second batch, this time from the beginning of biotransformation. On the other hand, the low value of DO provides a prerequisite, indicating that the AMX consortium was only temporarily suppressed and not inhibited, presented by production of gaseous nitrogen and formation of ions N–$NO_3^-$. The NR and NRRs achieved were similar to the results of the first batch, owing to increasing nitrite production.

### 2.1. *Influence of pH on the Process*

After the second biotransformation, the operation conditions were changed, which were intended to provide an advantage to the AMX part of the immobilized biomass. The first parameter modified was the pH of the process, with reference to published data of pH optimum of 6.5–8.8 for AMX cultivation [7]. According to this fact, a spontaneous decrease of the pH value during the process was performed instead of a pH stat at 8.0, as previously used. Following this change, a decrease in pH from 8.0 to 6.18 (after 122 h) and from pH 7.6 to 6.0 (after 110 h) was observed in the third and fourth biotransformations, respectively. Although there was no change in nitrite production during these processes, an increase in the utilization of N–$NH_4^+$ was observed, according to time of biotransformation. For the third batch, an NRR of 0.50 mg N (L h)$^{-1}$ and 42% NR was achieved, whereas for the fourth batch, the NRR was 0.59 mg N (L h)$^{-1}$ and 43% NR was achieved. Compared with previously achieved values, and as illustrated in Figure 2, there is a clear increase in the NRR despite a decrease in NR, which represents the activity of the whole process. During the fifth biotransformation, the pH stat was reset to pH 7.0 and, as shown in Figure 1, the increase in N–$NO_2^-$ concentration was significantly reduced for only 4.4 mg N L$^{-1}$ in the period 0–120 h, as compared with previous (second to fourth) conversions with an increase of about 31.9 ± 0.3 mg N L$^{-1}$.

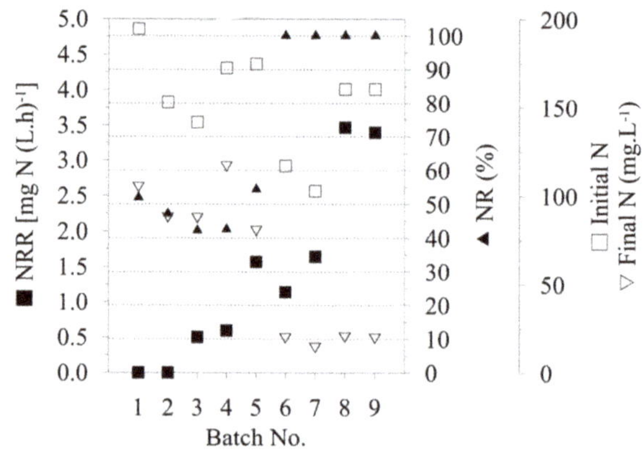


**Figure 2.** Nitrogen removal rate (NRR) calculated for the period 0–30 h of each biotransformation cycle, nitrogen removal (NR) calculated for entire duration of biotransformation and total, initial and final nitrogen.

## 2.2. *AMX Bacteria Stimulation*

Compared to AOB, AMX bacteria has extremely slow biomass growth (doubling time between 2.1 and 3.9 days) [18]. Therefore, AMX biomass suffers to be overgrown by AOB in the previous repeated batches. To increase the AMX bacterium, the nutritional limitation (decrease of N–$NH_4^+$) of AOB was applied. In the following two biotransformations (sixth and seventh), the initial concentration of N–$NH_4^+$ was decreased two-fold in an attempt to stimulate AMX bacteria at the expense of the other co-immobilized bacteria. At first, total utilization of N–$NH_4^+$ was observed after 25 h of biotransformation (sixth batch). Then, after 100 h, relatively high-speed utilization of N–$NO_2^-$ was observed, which decreased a concentration to 14.2 mg N L$^{-1}$ from 78.1 mg N L$^{-1}$ after 45 h (145th hour of biotransformation) by simultaneous formation of 19.52 mg N L$^{-1}$ (N–$NO_3^-$). The N–$NO_3^-$ confirms the trend of AMX bacteria activation at the end of this batch. Finally, a total NR of 100% was observed (calculated on initial N–$NH_4^+$, N–$NO_2^-$) despite it being at the expense of an NRR decrease (Figure 2). During the seventh batch, synchronic utilization of both sources of nitrogen was observed from the beginning of the biotransformation. An NRR of 1.65 mg N (L h)$^{-1}$ was achieved, as calculated for the period of 0–30 h, which is similar to the value of 1.57 mg N (L h)$^{-1}$ that was achieved in the fifth biotransformation, with the difference of 100% NR observed during seventh biotransformation (Figure 2). This confirmed the correctness of the chosen procedure in the sixth batch, with a decrease in the initial concentrations of N–$NH_4^+$ for AMX stimulation, and led to improved total nitrogen utilization and activity of the whole process with immobilized co-cultures. Furthermore, the effectivity of the process was also apparent from the initial and final content of total N between sixth and ninth batch (Figure 2). The difference between initial and final N content confirms the efficient process strategy in batch numbers eight and nine.

### 2.3. Harmonised Co-Culture Experiments

During the eighth and ninth biotransformations, the initial concentration of N–$NH_4^+$ was increased to 100.3 and 102.3 mg N $L^{-1}$, respectively. The initial concentration of N–$NO_2^-$ was similar to that used in previous repeated batch biotransformations at 60.1 and 58.1 mg N $L^{-1}$ for the eighth and ninth biotransformations, respectively. As shown in Figure 1, simultaneous utilization of N–$NH_4^+$ and N–$NO_2^-$ was observed from the beginning of both biotransformations. For both, 100% NR was observed after 75 h (ninth batch) and 78 h (eighth batch).

During the eighth batch, the higher NRR of 3.46 mg N $(L\ h)^{-1}$ was achieved in the 0–30 h period (Figure 2), and a similar rate of 3.39 mg N $(L\ h)^{-1}$ was observed during the ninth biotransformation. Although the highest achieved value of the NRR was 2.8 times lower than published data on a similar topic [9.8 mg N $(L\ h)^{-1}$], the observed NR efficiency in this study achieved 100% compared to reported 85% to 95% by Jenni et al. [12]. However, in this study, only small amount of immobilized biomass was applied (10% w/v particles load; in total 0,6 g of co-immobilized biomass per L of treated waste water). As reported recently [19], further process intensification is possible by the increase of immobilized biomass up to 20 g $L^{-1}$ of dry cell weight biomass, with no change in immobilized particles structure. Interestingly, this overload of biomass results in linear increase of the specific immobilized biomass activity as well. In the case of AMX and AOB co-immobilization, this may result in a more than 30-fold increase of initial specific activity. The further increase in the NRR can be also obtained by increase of the total amount of used immobilized particles (10% w/v in these experiments) to maximum of 30% (w/v) [20], which results in an increase of total activity of the biotransformation system. Therefore, we believe that this co-immobilized system has a potential for further applications in waste water treatment processes.

## 3. Materials and Methods

### 3.1. Medium

The cultivation medium (0.8 L) was composed of real wastewater (received as a product from machine-sludge thickening) obtained from Central WWTP Vrakuňa, Bratislava, Slovakia, and aged tap water (left to sit for 24 h for chlorine evaporation) in a volumetric ratio of 3:5. The characteristics of the wastewater were a pH value of 7.4, a chemical oxygen demand of 370 mg $L^{-1}$, a five-day biochemical oxygen demand of 138 mg $L^{-1}$, a suspended solids content of 678 mg $L^{-1}$, an ammonium nitrogen (N–$NH_4^+$) concentration of 275 mg $L^{-1}$ and a total nitrogen content of 673 mg $L^{-1}$. The nitrite nitrogen (N–$NO_2^-$) and nitrate nitrogen (N–$NO_3^-$) contents were not detectable. To increase the N–$NO_2^-$ concentration, 0.2 g of $NaNO_2$ (Mikrochem, Slovakia) was added into the cultivation medium. The final concentration of N–$NH_4^+$ in the cultivation medium was 103 ± 2 mg N $L^{-1}$, except for the sixth and seventh conversions, which started with concentrations of 38.8 and 45.8 mg N $L^{-1}$, respectively. The final concentration of N–$NO_2^-$ was 62 ± 15 mg $L^{-1}$. The medium was purged with $N_2$ for 15 min to remove the dissolved $O_2$, but this was only performed before the first biotransformation.

### 3.2. Biomass Immobilisation

Biomass used for immobilization was received from the Institute of Chemical Technology (ICT), Prague, Czech Republic, propagated in lab-scale conditions by use of real wastewater obtained from WWTP Prague. Using fluorescence in situ hybridization (FISH) by ICT, the composition of the received consortium of bacteria was characterized as 20% to 30% AMX bacteria and 70% to 80% of AOB. The biomass (1 ± 0.2 g wet weight) was softly suspended in 10 mL of aged tap water and mixed with 190 mL PVA hydrogel, prepared by using of PVA (20 g) and polyethylene glycol (12 g), which were melted in deionized water (158 mL) at 90 °C until clarification and then cooled to 30 °C. Lens-shaped gel particles with entrapped biomass were prepared by passing the gel mixture through thin nozzles to a hard surface, followed by subsequent drying in an airflow cabinet for 55 min. Particles were dried down to 30% of their initial mass and swollen in a stabilizing solution of 0.1 M sodium sulfate for

30–45 min. Then, the particles were separated through a sieve and washed from the stabilizing solution by deionized water. Using the mentioned LentiKats® method of immobilization, a final amount of 85 g of immobilized biomass was acquired.

### 3.3. Repeated Batch Biotransformation

Batch biotransformations with immobilized biomass were performed in a 1.3 L BioFlo® 115 fermenter (New Brunswick, USA) in which 0.8 L of the cultivation medium was inoculated with 85 g of LentiKats®. Each biotransformation was carried out at either pH 8.0 (first and second batch) or pH 7.0 (fifth to ninth batch) by automatic addition of 2 M NaOH at 30 °C with 250 rpm stirring. Repeated batch-mode biotransformations occurred when the residual concentrations of $N–NH_4^+$ and $N–NO_2^-$ were reduced to almost 0 mg N $L^{-1}$ or did not change in the long term. This situation was solved by separating the entire volume of the production medium through a sieve to avoid washing out the LentiKats® with the immobilized biomass. The fermenter was subsequently fed with a fresh cultivation medium. Experiments were duplicated, and datapoints represent mean value of the process (SD was lower than 5%).

### 3.4. Analytical Assays

The concentrations of $N–NH_4^+$, $N–NO_2^-$ and $N–NO_3^-$ were determined by methods described in the Standard Methods for the Examination of Water and Wastewater [21], using a BioSpectometer® (Eppendorf, Germany). Dissolved oxygen (DO) was measured using a polarographic DO probe, and the pH value of the medium was measured using a pH probe, both of which are components of the BioFlo® 115 fermenter. NR and NRR were calculated to the depletion of primary sources of nitrogen: $N–NH_4^+$ and $N–NO_2^-$.

## 4. Conclusions

The presented results emphasize that the use of a small amount of immobilized biomass (0.6 g of immobilized wet weight per L of treated waste water) can be advantageous in large-scale applications, owing to the low biomass-production rate of AMX bacteria and AOB. The decrease of $N–NH_4^+$ leads to starvation of AOB, which stimulates the AMX bacteria and significantly improves the total nitrogen utilization and activity of co-immobilized cultures. This starvation technique is a useful trick for co-culture activities harmonization, which resulted in high overall nitrogen removal activity. At optimal conditions pH = 7 and initial concentrations $N–NH_4^+$ 100.3 mg N $L^{-1}$ and $N–NO_2^-$ 60.1 mg $L^{-1}$, at 30 °C with gentle (250 rpm) batch bioreactor stirring, an immobilized consortium was able to remove 160.4 mg N $L^{-1}$ of initial nitrogen sources from wastewater within 80 h with NR rate 3.46 mg N $(L\ h)^{-1}$. This is the first report on the immobilization of AMX bacteria and AOB into a PVA hydrogel to indicate the methods, pH stat and substrate limitation that stimulate the co-immobilized bacteria activity in biotransformations. This co-culture improvement strategy might be beneficial for further co-immobilization studies and applications.

**Author Contributions:** I.D. investigation, writing—original draft preparation; R.S. methodology; M.R. (Michal Rosenberg) conceptualization, funding acquisition; M.R. (Martin Rebroš) writing—review and editing, methodology, funding acquisition.

**Funding:** This work was supported by the Ministry of Industry and Trade of the Czech Republic, programme TIP, grant no. FR-TI4/254 and was cofounded by the Slovak Research and Development Agency under contract no. APVV-16-0314. This publication is the result of the project implementation: Comenius University in Bratislava Science Park supported by the Research and Development Operational Programme funded by the ERDF. Grant number: ITMS 26240220086.

**Conflicts of Interest:** The authors declare no conflict of interest.

## References


1. Ni, B.-J.; Smets, B.F.; Yuan, Z.; Pellicer-Nàcher, C. Model-based evaluation of the role of Anammox on nitric oxide and nitrous oxide productions in membrane aerated biofilm reactor. *J. Membr. Sci.* **2013**, *446*, 332–340. [CrossRef]
2. Paredes, D.; Kuschk, P.; Mbwette, T.S.A.; Stange, F.; Müller, R.A.; Köser, H. New aspects of microbial nitrogen transformations in the context of wastewater treatment—A review. *Eng. Life Sci.* **2007**, *7*, 13–25. [CrossRef]
3. Bi, Z.; Qiao, S.; Zhou, J.; Tang, X.; Cheng, Y. Inhibition and recovery of Anammox biomass subjected to short-term exposure of Cd, Ag, Hg and Pb. *Chem. Eng. J.* **2014**, *244*, 89–96. [CrossRef]
4. Liu, S.; Zhang, Z.; Ni, J. Effects of $Ca^{2+}$ on activity restoration of the damaged anammox consortium. *Bioresour. Technol.* **2013**, *143*, 315–321. [CrossRef] [PubMed]
5. Ma, B.; Peng, Y.; Zhang, S.; Wang, J.; Gan, Y.; Chang, J.; Wang, S.; Wang, S.; Zhu, G. Performance of anammox UASB reactor treating low strength wastewater under moderate and low temperatures. *Bioresour. Technol.* **2013**, *129*, 606–611. [CrossRef] [PubMed]
6. Gao, F.; Zhang, H.; Yang, F.; Qiang, H.; Li, H.; Zhang, R. Study of an innovative anaerobic (A)/oxic (O)/anaerobic (A) bioreactor based on denitrification–anammox technology treating low C/N municipal sewage. *Chem. Eng. J.* **2013**, *232*, 65–73. [CrossRef]
7. Magrí, A.; Béline, F.; Dabert, P. Feasibility and interest of the anammox process as treatment alternative for anaerobic digester supernatants in manure processing—An overview. *J. Environ. Manag.* **2013**, *131*, 170–184. [CrossRef] [PubMed]
8. Huang, X.-W.; Wei, Q.-Y.; Urata, K.; Tomoshige, Y.; Zhang, X.-H.; Kawagoshi, Y. Kinetic study on nitrogen removal performance in marine anammox bacterial culture. *J. Biosci. Bioeng.* **2014**, *117*, 285–291. [CrossRef] [PubMed]
9. Jin, R.-C.; Xing, B.-S.; Yu, J.-J.; Qin, T.-Y.; Chen, S.-X. The importance of the substrate ratio in the operation of the Anammox process in upflow biofilter. *Ecol. Eng.* **2013**, *53*, 130–137. [CrossRef]
10. Isaka, K.; Kimura, Y.; Yamamoto, T.; Osaka, T.; Tsuneda, S. Complete autotrophic denitrification in a single reactor using nitritation and anammox gel carriers. *Bioresour. Technol.* **2013**, *147*, 96–101. [CrossRef] [PubMed]
11. Anjali, G.; Sabumon, P.C. Unprecedented development of anammox in presence of organic carbon using seed biomass from a tannery Common Effluent Treatment Plant (CETP). *Bioresour. Technol.* **2017**, *153*, 30–38. [CrossRef] [PubMed]
12. Jenni, S.; Vlaeminck, S.E.; Morgenroth, E.; Udert, K.M. Successful application of nitritation/anammox to wastewater with elevated organic carbon to ammonia ratios. *Water Res.* **2014**, *49*, 316–326. [CrossRef] [PubMed]
13. Magrí, A.; Vanotti, M.B.; Szögi, A.A. Anammox sludge immobilized in polyvinyl alcohol (PVA) cryogel carriers. *Bioresour. Technol.* **2012**, *114*, 231–240. [CrossRef] [PubMed]
14. Furukawa, K.; Inatomi, Y.; Qiao, S.; Quan, L.; Yamamoto, T.; Isaka, K.; Sumino, T. Innovative treatment system for digester liquor using anammox process. *Bioresour. Technol.* **2009**, *100*, 5437–5443. [CrossRef] [PubMed]
15. Isaka, K.; Date, Y.; Sumino, T.; Tsuneda, S. Ammonium removal performance of anaerobic ammonium-oxidizing bacteria immobilized in polyethylene glycol gel carrier. *Appl. Microbiol. Biotechnol.* **2007**, *76*, 1457–1465. [CrossRef] [PubMed]
16. Rebroš, M.; Rosenberg, M.; Stloukal, R.; Krištofíková, L. High efficiency ethanol fermentation by entrapment of Zymomonas mobilis into LentiKatsR. *Lett. Appl. Microbiol.* **2005**, *41*, 412–416. [CrossRef] [PubMed]
17. Qiao, S.; Tian, T.; Duan, X.; Zhou, J.; Cheng, J. Novel single-stage autotrophic nitrogen removal via co-immobilizing partial nitrifying and anammox biomass. *Chem. Eng. J.* **2013**, *230*, 19–26. [CrossRef]
18. Zhang, L.; Narita, Y.; Gao, L.; Ali, M.; Oshiki, M.; Okabe, S. Maximum specific growth rate of anammox bacteria revisited. *Water Res.* **2017**, *116*, 296–303. [CrossRef] [PubMed]
19. Zajkoska, P.; Rosenberg, M.; Heath, R.; Malone, K.J.; Stloukal, R.; Turner, N.J.; Rebros, M. Immobilised whole-cell recombinant monoamine oxidase biocatalysis. *Appl. Microbio. Biotechnol.* **2015**, *99*, 1229–1236. [CrossRef] [PubMed]

20. Stloukal, R.; Rosenberg, M.; Rebros, M. Method for industrial production of biocatalysts in the form of enzymes or microorganisms immobilized in polyvinyl alcohol gel, their use and devices for their production. WO/2007/104268, 20 December 2007.
21. APHA, AWWA, WEF. *Standard Methods for the Examination of Water and Wastewater*, 21st ed.; American Public Health Association/American Water Works Association/Water Environment Federation: Washington DC, WA, USA, 1998.

catalysts

MDPI

Article

# Expression and Characterization of a Dye-Decolorizing Peroxidase from *Pseudomonas Fluorescens* Pf0-1

**Nikola Lončar [1,2,*], Natalija Drašković [3], Nataša Božić [4], Elvira Romero [2,†], Stefan Simić [3], Igor Opsenica [3], Zoran Vujčić [3] and Marco W. Fraaije [2]**

1 GECCO Biotech, Nijenborgh 4, 9747AG Groningen, The Netherlands
2 Molecular Enzymology group, University of Groningen, Nijenborgh 4, 9747AG Groningen, The Netherlands; elvira.romero@astrazeneca.com (E.R.); m.w.fraaije@rug.nl (M.W.F.)
3 Faculty of Chemistry, University of Belgrade, Studentski trg 12-16, 11000 Belgrade, Serbia; draskovic.natalija@gmail.com (N.D.); ssimic@chem.bg.ac.rs (S.S.); igorop@chem.bg.ac.rs (I.O.); zvujcic@chem.bg.ac.rs (Z.V.)
4 ICTM-Center of Chemistry, University of Belgrade, Studentski trg 12-16, 11000 Belgrade, Serbia; nbozic@chem.bg.ac.rs
* Correspondence: n.loncar@rug.nl
† Current address: AstraZeneca R&D Gothenburg, Discovery Sciences, S-431 83 Mölndal, Sweden.

Received: 7 April 2019; Accepted: 16 May 2019; Published: 20 May 2019

**Abstract:** The consumption of dyes is increasing worldwide in line with the increase of population and demand for clothes and other colored products. However, the efficiency of dyeing processes is still poor and results in large amounts of colored effluents. It is desired to develop a portfolio of enzymes which can be used for the treatment of colored wastewaters. Herein, we used genome sequence information to discover a dye-decolorizing peroxidase (DyP) from *Pseudomonas fluorescens* Pf-01. Two genes putatively encoding for DyPs were identified in the respective genome and cloned for expression in *Escherichia coli*, of which one (*Pf*DyP B2) could be overexpressed as a soluble protein. *Pf*DyP B2 shows some typical features known for DyPs which includes the ability to convert dyes at the expense of hydrogen peroxide. Interestingly, *t*-butyl hydroperoxide could be used as an alternative substrate to hydrogen peroxide. Immobilization of *Pf*DyP B2 in calcium-alginate beads resulted in a significant increase in stability: *Pf*DyP B2 retains 80% of its initial activity after 2 h incubation at 50 °C, while the soluble enzyme is inactivated within minutes. *Pf*DyP B2 was also tested with aniline and ethyl diazoacetate as substrates. Based on GC-MS analyses, 30% conversion of the starting material was achieved after 65 h at 30 °C. Importantly, this is the first report of a DyP-catalyzed insertion of a carbene into an N-H bond.

**Keywords:** DyP peroxidase; oxidoreductase; reactive dye; decolorization

## 1. Introduction

Peroxidases are oxidoreductases involved in a variety of biochemical processes, including the biosynthesis of cell wall material and immunological host-defense responses [1]. The recently discovered DyP-type peroxidases (DyPs, dye-decolorizing peroxidases; EC 1.11.1.19) represent a novel superfamily of heme-containing enzymes. They share no significant similarity in primary sequence or structure to other peroxidase superfamilies [2]. DyPs possess a broad substrate specificity and low pH optimum [2], while they are most stable at neutral pH [3]. Using hydrogen peroxide as an electron acceptor, they are capable of catalyzing efficient oxidations of a wide array of industrially relevant substrates including dyes with anthraquinone structure, β-carotene, and aromatic sulfides [4–8]. Instead of a distal histidine acting as an acid-base catalyst, DyPs contain an aspartate active site residue within a highly conserved GXXDG distal motif [8], although recent findings indicate that instead of aspartate some DyPs have a glutamate [9]. On the proximal side, the heme iron of DyPs is coordinated by a histidine.

Several bacterial DyPs were found to be robust enzymes and thus potent biocatalysts [10–12]. Since they can degrade a variety of synthetic dyes, DyPs can potentially be used in the bioremediation of dye-contaminated waste water. Two fungal DyP-type peroxidases were shown to degrade $\beta$-carotene [7], which is of tremendous interest in the food industry for enabling the enzymatic whitening of whey-containing foods and beverages. DyP peroxidases also show promise as novel anti-microbial (pro)drug targets [13]. Their myriad industrial uses point to the importance of discovery and characterization of these enzymes.

In this work, we have used the available genome sequence of *Pseudomonas fluorescens* Pf0-1 to identify a gene coding for a DyP (*Pf*DyP B2). This peroxidase has been overexpressed in *Escherichia coli* and purified. It shows activity towards various substrates including 2,6-dimethoxyphenol, 2,2′-Azino-bis(3-ethylbenzothiazoline-6-sulfonic acid) (ABTS), *o*-dianisidine, and reactive dyes. Besides hydrogen peroxide, *Pf*DyP B2 was able to use *t*-butyl hydroperoxide as an electron acceptor. *Pf*DyP B2 was also probed for a reaction that was recently reported to be catalyzed by several heme-containing proteins including cytochrome P450 monooxygenases and myoglobin: carbonyl-olefination, cyclopropanation, and N/S-H insertion reactions using the reaction of ethyl diazoacetate with heme under anaerobic conditions [14–17]. Recently, it was also shown that a DyP, YfeX from *E. coli*, is able to perform carbonyl olefination [18]. YfeX was initially reported to be a deferrochelatase, extracting iron ions from heme [19]. However, it was later identified as a DyP [20,21]. Another recent study on the implementation of YfeX in the synthesis of tryptamine precursors [22] triggered our attention to investigate whether *Pf*DyP B2 can be similarly used as a hemoprotein scaffold for reactions not probed before using a DyP. Our work shows that the reaction between aniline and ethyl diazoacetate can be satisfactorily achieved using *Pf*DyP B2 as a biocatalyst. Remarkably, this is the first report of a DyP-catalyzed insertion of a carbene into an N-H bond.

## 2. Results

### 2.1. Expression and Purification

A search for DyP peroxidases in the predicted proteome of *P. fluorescences* Pf-01 using RedoxiBase (http://peroxibase.toulouse.inra.fr/) revealed that this organism contains three peroxidases that belong to the DyP peroxidase superfamily, one of which belongs to clade A DyPs and two to clade B DyPs. In addition, one gene encoding a catalase-peroxidase (KatG) was identified. Clade A DyPs have been hypothesized to be less catalytically active due to a mixed spin state of the heme [23]. Two genes encoding for DyPs belonging to clade B were amplified by PCR using genomic DNA of *P. fluorescens* Pf-01 as a template. These fragments have been cloned into a pBAD vector and expression was carried out under different conditions. *Pf*DyP B1 could not be expressed in sufficient amounts (data not shown), while *Pf*DyP B2 was overexpressed and appeared as the most dominant protein band upon SDS-PAGE analysis of the cell-free extract (Figure 1a). The predicted molecular weight of *Pf*DyP B2 is 35 kDa. Purification of the cell free extract using $Ni^{2+}$-Sepharose showed that the binding of *Pf*DyP B2 to the resin was very poor since it eluted with less than 10 mM imidazole. This eluted fraction contained additional proteins hampering the biochemical characterization of PfDyP B2. Thus, further purification was performed. Mixed mode resins enable the purification of biomolecules without feedstream conditioning and can operate across a wide range of salt concentrations and pH. One of the mixed mode resins is Nuvia aPrime 4A (Bio Rad), which is designed with a positively charged hydrophobic ligand, which helps in biomolecule binding by both hydrophobic and anionic interactions. Therefore, Nuvia aPrime 4A mixed mode resin was used for polished purification of *Pf*DyP B2. Active peroxidase eluted with 1 M NaCl and showed ~95% purity upon SDS-PAGE analysis (Figure 1).

The purification procedure yielded ~35 mg of purified His-tagged *Pf*DyP B2 from 1 L of culture broth. This two-step workflow is readily scalable for process production.

Previous studies have shown that DyPs represent a novel peroxidase superfamily harboring a heme molecule in their structure [24,25]. A UV-Vis absorption spectrum of the purified enzyme was recorded between 250 and 750 nm. The Soret band at 405 nm (Figure 1b) indicates that the heme cofactor had indeed been incorporated into the purified PfDyP B2. The Reinheitzahl value (Rz, the ratio of $A_{405}/A_{280}$) for the purified enzyme was 1.16.

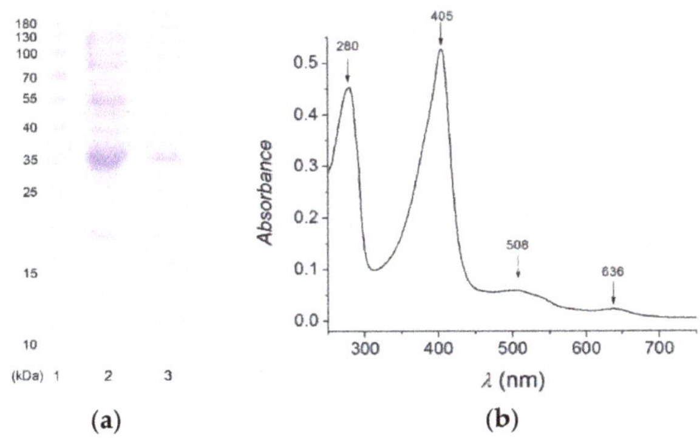


**Figure 1.** (**a**) SDS PAGE analysis. Lane 1: Molecular markers; lane 2: Cell free extract; lane 3: *Pf*DyP B2 fraction after purification with Nuvia aPrime 4A chromatography. (**b**) UV-Vis spectrum of *Pf*DyP B2.

## 2.2. Biochemical Characterization

The purified enzyme was most active at pH 4.0 (Figure 2) with moderate activity between pH 3.0 and 6.0, which is in perfect agreement with other studies on DyPs [9,23,26].

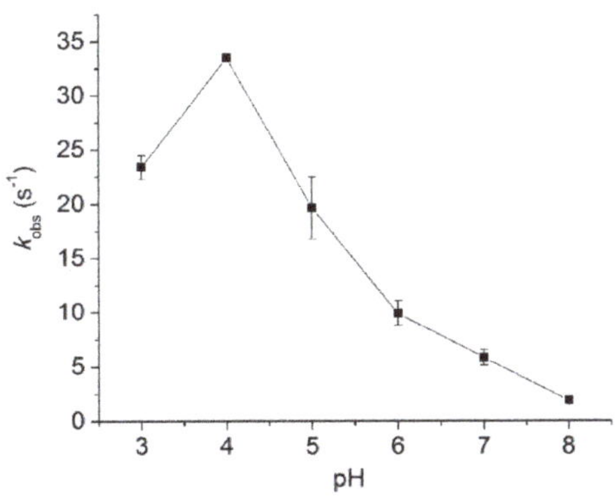


**Figure 2.** The effect of pH on the observed rate ($k_{obs}$) of oxidation of ABTS by *Pf*DyP B2.

The relative activity of the purified *Pf*DyP B2 enzyme was tested using 0.5 mM ABTS and 0.1 mM hydrogen peroxide in the presence of various cations ($Ca^{2+}$, $Mg^{2+}$, $Zn^{2+}$, $Mn^{2+}$, $Co^{2+}$, $Fe^{2+}$, and $Hg^{2+}$) and reducing agents (aminotriazole, EDTA, imidazole, DTT, Cys, and Na-azide). The results show that the relative activity was not diminished by the addition of Ca, Mg, Zn, Mn, and Co ions, but was significantly reduced in the presence of Fe and Hg ions (Figure 3a). Partial inhibition of the purified enzyme was observed with aminotriazole, EDTA, and imidazole, whereas almost complete inhibition was seen with DTT, Cys, and Na-azide (Figure 3b).

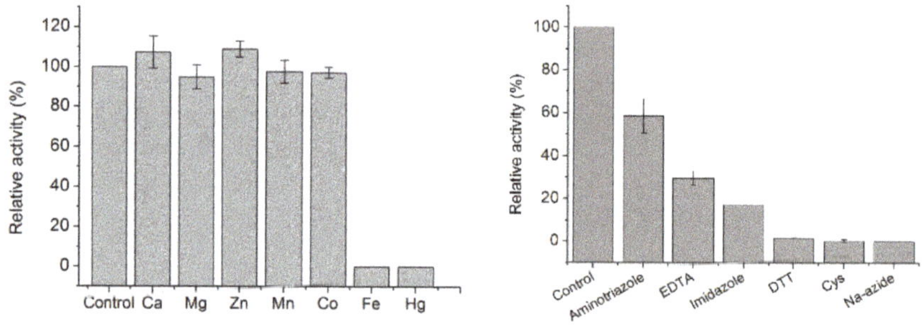


(**a**) (**b**)

**Figure 3.** (**a**) The effect of metal ions on activity of *Pf*DyP B2. (**b**) The effect of inhibitors on the activity of *Pf*DyP B2. Activity in the absence of inhibitors was taken as 100%.

To investigate the substrate scope of the purified *Pf*DyP B2, it was tested using various well-known peroxidase substrates at ambient temperature. As shown in Table 1, *Pf*DyP B2 showed activity towards most of the assayed substrates including ABTS and aromatic, azo, and anthraquinone dyes. In addition, guaiacol, 2,6-dimethoxyphenol, manganese, veratryl alcohol, CBZ-ethanolamine, syringaldehyde, and acetosyringone were tested, but *Pf*DyP B2 was inactive towards them. The reactions were carried out in 50 mM Na-acetate buffer pH 4.0 for ABTS, *o*-dianisidine, and pyrogallol. In the case of reactive dyes, assays were carried out using 50 mM sodium-acetate buffer at pH 3.0 since a tenfold higher activity towards these dyes was observed at this pH value.

**Table 1.** Kinetic parameters measured for *Pf*DyP B2.

| Substrate | $\lambda$ (nm) | $\varepsilon$ ($mM^{-1}cm^{-1}$) | $K_m$ (mM) | $k_{cat}$ ($s^{-1}$) |
|---|---|---|---|---|
| ABTS | 414 | 36.6 | 0.22 ± 0.04 | 102 ± 6.2 |
| *o*-Dianisidine | 460 | 11.3 | 0.003 ± 0.001 | 14.7 ± 1.7 |
| Pyrogallol | 430 | 2.47 | 1.45 ± 0.63 | 1.98 ± 0.03 |
| Reactive blue 4 | 610 | 4.2 | 0.010 ± 0.004 | 1.54 ± 0.13 |
| Reactive black 5 | 597 | 37 | 0.006 ± 0.002 | 0.04 ± 0.01 |
| $H_2O_2$ [a] | 240 | 0.0394 | 0.52 ± 0.12 | 23 ± 0.84 |
| *t*-BuOOH [a] | n.a. | n.a. | 415 ± 144 | 97 ± 28 |

[a] Measured by using ABTS as substrate. n.a.- not applicable

Using Kraft lignin as a substrate, a Michaelis-Menten saturation curve was observed (Figure 4), similar to that previously observed for other DyPs [27,28].

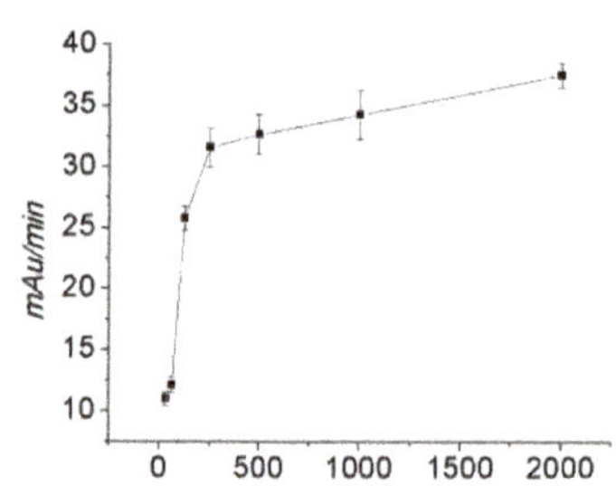


**Figure 4.** The steady-state activity of *Pf*DyP B2 with Kraft lignin as monitored by absorbance change at 465 nm [27,28].

*Pf*DyP B2 showed poor stability at 40 and 50 °C, being inactivated within minutes. However, its activity remained unaffected for at least 2 h at 30 °C. Among existing strategies for the improvement of enzyme stability, a well-known approach is enzyme immobilization. Immobilization also offers the possibility to reuse the respective enzyme and allows easy separation of enzymes and product, which are factors influencing the overall cost of enzymatic industrial processes. *Pf*DyP B2 was immobilized in alginate by mixing an enzyme solution with 2% sodium-alginate followed by droplet ejection into a 2% $CaCl_2$ solution. The beads were left to age in the $CaCl_2$ solution followed by washing steps with buffer. Fifty percent of the enzyme leaked out of the beads according to activity and protein measurements. The remaining enzyme was retained in the beads. Immobilized *Pf*DyP B2 retained 80% of initial activity after a 2 h incubation at 50 °C, revealing a significant increase in stability compared to soluble enzyme.

Encouraged by previous reports of heme-containing enzymes catalyzing the insertion of carbenes into N-H bonds [15,17], a reaction between aniline and ethyl diazoacetate was investigated using *Pf*DyP B2 as a catalyst (Scheme 1). Aniline and ethyl diazoacetate were reacted in the presence of sodium dithionite as a reductant and 0.4 mol% enzyme, using acetate buffer, pH 4.5. The reaction was kept at 30 °C for 65 h. Subsequently, the product formation was analyzed and confirmed by GC-MS. Conversion of the starting material was determined to be 30%.

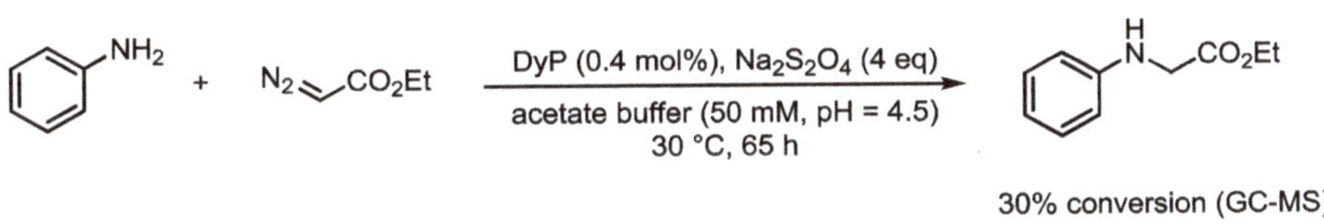


**Scheme 1.** The *Pf*DyP B2-catalyzed reaction between aniline and ethyl diazoacetate.

## 3. Discussion

Dye-decolorizing peroxidases can be useful biocatalysts for industrial applications due to their extraordinary ability to degrade a variety of synthetic dyes. They have the potential to be used for the bioremediation of dye-contaminated waste water. In the current study, we have purified a His-tagged DyP from *P. fluorescens* Pf-01 expressed in *E. coli*. Purification was performed on an IMAC resin which resulted in a partially pure target protein. A hydrophobic anion exchanger, Nuvia aPrime 4A, was used to obtain highly pure PfDyP B2. Although it was not tested in this particular instance, it may be that Nuvia aPrime 4A can provide a single-step purification directly from the crude cell extract. Conveniently this resin is compatible with high salt concentration and a broad pH range without the need for feed dilution/reconstitution. This would help in limiting the number of steps required to generate a highly pure enzyme sample. While IMAC is widely used for purification of recombinant proteins, mixed-mode resins should also be considered when it is not so straightforward to achieve a desired purity in one step by affinity chromatography.

The steady-state kinetic parameters determined for PfDyP B2 indicate that it is highly active on ABTS and *o*-dianisidine. This suggests that PfDyP B2 may be exploited for the biomedical applications and various assays which are typically based on the prototype plant peroxidase, horseradish peroxidase (HRP). Reactive blue 4 is oxidized by PfDyP B2 at a similar rate to pyrogallol, while Reactive black 5 is a more recalcitrant dye (Table 1).

Inactivation by hydrogen peroxide is often underlined as the weakness of peroxidases when it comes to industrial application. However, it is also known that this problem can be solved by using alternative electron acceptors [29]. Interestingly, beside hydrogen peroxide, *Pf*DyP B2 was able to use *t*-butyl hydroperoxide as an electron acceptor. The $k_{cat}$ value using this compound is four-fold higher than when using hydrogen peroxide. However, the $K_m$ for *t*-butyl hydroperoxide is quite high which is limiting the possibility of using it in wastewater treatment. While *t*-butyl hydroperoxide is a more expensive electron acceptor than hydrogen peroxide, it may still offer some advantages. The use of *t*-butyl hydroperoxide can reduce the overall enzyme loading in the process since enzyme inactivation by hydrogen peroxide will be minimized. This is something that would have to be optimized case by case though reaction engineering and is process-dependent. In large-scale industrial processes, such as dye decolorization, the use of *t*-butyl hydroperoxide is probably not feasible. However, DyPs may also be used in some other applications, for example for the production of high-value chemicals, as described for *Tfu*DyP for enantioselective sulfoxidation [8]. The acceptance of *t*-butyl hydroperoxide by the enzyme also discloses mechanistic information; apparently, this organic oxidant is able to reach the heme while this is not the case for all peroxidases.

Prior studies have also shown that DyP-type peroxidases can act on lignin and can be involved in its degradation [30]. Enzymatic lignin depolymerization is highly attractive, as it offers access to a wide array of aromatic precursors which can be further functionalized for the production of high-added value chemicals. We tested if *Pf*DyP B2 can act on Kraft lignin. As shown in Figure 4, we detected an increased DyP activity with an increase in the Kraft lignin concentration. Other DyPs acting on lignin and originating from *Pseudomonas* sp. are described in the literature, giving further confirmation of the potential of *Pf*DyP B2 for lignin valorization [31]. One must keep in mind that the lignin depolymerization process is very complicated and will hardly be achieved by one enzyme. Considering the radical mechanism of laccases and peroxidases it is certain that the concerted action of several enzymes or trapping of produced radicals is needed to achieve depolymerization. Otherwise, depolymerization and repolymerization will occur, as demonstrated in recent publications of synthesis of lignin oligomers [11,32].

Any scale up of enzymatic processes is still highly dependent on the cost of enzymes. Joint efforts of increasing the expression levels and the efficient use of enzymes are needed to make the process economically feasible. Our first attempts at the immobilization of *Pf*DyP B2 have demonstrated that this approach is clearly beneficial. The immobilized enzyme was able to withstand increased temperature as compared to its soluble counterpart. Magnetic particles may further aid and improve the activity of DyPs as shown recently for a similar DyP from *P. fluorescens* Pf-5 [33].

Except for valorising lignin by using DyPs, these heme-containing enzymes may also be used for non-natural reactions. Recent papers have shown that hemoproteins can be used for a number of attractive reactions [14–17]. This study shows that *Pf*DyP B2 can be used to perform an insertion of a carbene in an N-H bond. To our knowledge, this is the first time that a DyP was able to catalyze such a reaction.

## 4. Materials and Methods

### 4.1. Reagents and Enzymes

*Pfu* polymerase, restriction enzymes, and prestained protein ladders were from Thermo Scientific. $Ni^{2+}$-Sepharose HP was from GE Lifescience. All other chemicals were supplied by Sigma-Aldrich (St. Louis, MO, USA) and were of analytical grade.

### 4.2. Strains, Plasmids, and Growth Conditions

The *E. coli* strain TOP10 was used for routine cloning and maintenance of all plasmid constructs. This strain was also used for overexpression of *Pf*DyP B1 and *Pf*DyP B2. The genes encoding *Pf*DyP B1 (RefSeq: YP_348895.1) and *Pf*DyP B2 (RefSeq: YP_348987.1) were amplified by polymerase chain reaction from genomic DNA of *P. fluorescens* Pf0-1. The PCR fragments were amplified using primers GAC GTA <u>CAT ATG</u> AGT TAC TAC CAG C and ATA <u>GAA TTC</u> GCC GAT GCG CAG T for *Pf*DyP B1 and ACT <u>CAT ATG</u> TTG GGA GTC ACG ATC AT and ATT <u>GAA TTC</u> TCG CTC TGC CAA CTC TT for *Pf*DyP B2 cloning between NdeI and EcoRI restriction sites of the pBADNdeIHis vector to give the pBAD-*Pf*DyP B1 and pBAD-*Pf*DyP B2 constructs for the expression of peroxidases with an C-terminal mycHis tag. Restriction sites in primers are underlined. pBad*NdeI*His is a pBAD/Myc-HisA-derived expression vector (Thermo Fisher, Waltham, MA, USA) in which the *NdeI* site has been removed and the *NcoI* site has been replaced by *NdeI*. For verification, all created constructs have been sequenced (GATC Biotech).

### 4.3. Sequence Analysis

The ProtParam tool at the ExPASy server (URL: http://web.expasy.org/protparam/) was used to calculate extinction coefficients for *Pf*DyPB1 (36,440 $M^{-1}$ $cm^{-1}$) and *Pf*DyPB2 (25,690 $M^{-1}$ $cm^{-1}$).

### 4.4. Expression and Purification

*E. coli* TOP10 cells expressing *Pf*DyPB1 and *Pf*DyPB2 were grown in LB medium at 37 °C to saturation overnight. The following day, cultures were diluted 1:100 into fresh media and grown until $OD_{600}$ = 1.25 after which 0.02% L-arabinose and 0.75 mM hemin were added to induce the expression of *Pf*DyP B1 and B2, respectively. Expression was carried out at 30 °C, 180 rpm for 24 h. Cells were harvested by centrifugation at 6000 rpm and washed once with 50 mM potassium phosphate buffer pH 8.0 with 0.5 M NaCl. Pelleted cells were resuspended in the same buffer and disrupted by sonication. Cell-free extract was obtained after centrifugation at 11,000 rpm at 4 °C for 1 h.

For purification of His-tagged *Pf*DyP B2, cell free extract was loaded on a $Ni^{2+}$-Sepharose HP column equilibrated with 50 mM potassium phosphate buffer pH 8.0 with 0.5 M NaCl. Non-specifically bound proteins were washed away stepwise with 2 column volumes of 10 mM imidazole in starting buffer followed by elution with 250 mM imidazole in the same buffer. *P. fluorescens* DyP shows poor binding to the column but elutes as an observable red-colored fraction. Collected fractions were analyzed by reducing SDS-PAGE and subsequent protein staining. Fractions containing DyP were impure and were further purified on Nuvia aPrime 4A resin from Bio-Rad Laboratories. Binding to the resin was achieved using potassium phosphate buffer 50 mM pH 7.8 and purification was carried out with stepwise elution using starting buffer with 0.1 M NaCl, 0.2 M NaCl, 0.3 M NaCl, 0.5 M NaCl, and 1.0 M NaCl. Active peroxidase elutes with 1.0 M NaCl and shows high purity on SDS-PAGE analysis. Pure fractions were pooled and concentrated using an Amicon stirring cell equipped with a 10 kDa cut-off membrane.

Protein fractions were analyzed by 10% SDS-PAGE using the SE260 Mighty Small II Deluxe System. The UV-Vis absorption spectra of purified DyPB were recorded between 200 and 800 nm using a Shimadzu UV-1800 spectrophotometer in 1 cm quartz cuvettes at room temperature.

### 4.5. Steady-State Kinetic Analyses

DyP activity was measured spectrophotometrically using Shimadzu UV-1800 at ambient temperature. The pH optimum for activity was determined using 0.5 mM ABTS and 1.0 mM $H_2O_2$ using a set of buffers (0.10 M of Na-acetate pH 3.0–5.0, MES pH 6.0, TrisHCl 7.0–8.0, and glycine/NaOH pH 9.0–10.0). Upon determination of the optimal pH, further measurements were made in an appropriate buffer.

Kinetic parameters of peroxidase activity ($k_{cat}$ and $K_m$) were measured using different concentrations of ABTS, *o*-dianisidine, pyrogallol, Reactive blue 4, and Reactive black 5 with 1.0 mM $H_2O_2$ as co-substrate by adding 10 µL of suitably diluted enzyme (final concentration 5.3 nM). For $H_2O_2$ and *t*-butyl hydroperoxide, a fixed concentration of ABTS was used and the concentration of peroxide varied. Control reactions were included without using enzyme, $H_2O_2$, or both. Blanks were recorded in parallel with the measurements and subtracted accordingly. The kinetic parameters were calculated by fitting the data with a Michaelis-Menten equation using nonlinear analysis.

Kraft Lignin was tested as a substrate for *Pf*DyP B2 using a previously described procedure [27,28]. Briefly, Kraft lignin was dissolved in 0.10 M NaOH (10 mg/mL) and the pH was adjusted to 4.0 using 1.0 M sodium-acetate buffer. Different dilutions of this stock were mixed in 1 mL cuvette with *Pf*DyP B2 (25 µL, 1 mg/mL) and hydrogen peroxide (0.5 mM, final concentration), and the absorbance was monitored for 5 min at 465 nm. Inhibition of the peroxidase was tested using 3-amino-1,2,4-triazole, imidazole, DTT, cysteine, and sodium-azide. The enzyme was assayed for peroxidase activity as described above after preincubation with the inhibitor for 3 min prior to performing the assay with ABTS.

Thermal stability was determined by measuring residual activity upon incubating aliquots of *Pf*DyP B2 at 30 °C, 40 °C, and 50 °C. Samples were withdrawn at specific time points and activity was determined spectrophotometrically as described above.

### 4.6. Immobilization Study

*Pf*DyP B2 was immobilized in alginate by mixing 1 mL of enzyme solution (2 mg/mL) with 1 mL of 2% sodium-alginate followed by droplet ejection through a syringe needle into 1 L of 2% $CaCl_2$ solution, stirred using a mechanical overhead stirrer. The beads were left to age in the $CaCl_2$ solution followed by washing steps with buffer. The produced calcium alginate beads were collected and washed three times using 10 mL of 50 mM Na-acetate buffer pH 4.0. The leaking of *Pf*DyP B2 from alginate beads was checked in all wash fractions by measuring protein concentration using Bradford assay and activity of *Pf*DyP B2 using ABTS as substrate. The total protein amount in wash fractions was subtracted from the starting amount of *Pf*DyP B2 to estimate the amount of enzyme which was successfully entrapped in alginate beads. The immobilized enzyme was tested for activity using the ABTS assay before and after incubation in a water bath set at 50 °C.

### 4.7. Procedure for the DyP-Catalyzed Reaction Between Aniline and Ethyl Diazoacetate

The reaction was carried out in a 2 mL glass vial (Agilent Technologies, San Diego, CA, USA). To an unsealed vial charged with argon and a stirring bar, 840 μL of a 47 μM solution of DyP in acetate buffer (50 mM, pH = 4.5) was added. The vial was sealed with a silicone septum and the headspace of the vial was flushed with argon. A 1 M solution of sodium dithionite in acetate buffer (50 mM, pH 4.5) was degassed by bubbling with argon for 5 min, and 40 μL of this solution was added to the reaction mixture via a glass syringe. Solutions of ethyl diazoacetate and aniline were degassed in the same fashion prior to addition to the reaction mixture. Subsequently, 20 μL of a 0.5 M ethyl diazoacetate solution in a 1:1 (*v*/*v*) mixture of acetate buffer and dimethyl sulfoxide, and 100 μL of a 0.1 M solution of aniline in acetate buffer were added in the indicated order. The final concentrations of the added compounds were: 10 mM aniline, 10 mM ethyl diazoacetate, 40 mM sodium dithionite, and 40 μM DyP. The reaction mixture was stirred on a magnetic stirrer at 30 °C for 65 h. After the elapsed time, the mixture was extracted with ethyl acetate, dried over sodium sulfate, concentrated under reduced pressure, and analyzed by GC-MS.

GC-MS spectra (Supplementary Materials) of the synthesized compound were acquired on an Agilent Technologies 7890A apparatus equipped with a DB-5 MS column (30 m × 0.25 mm × 0.25 μm), a 5975C MSD and FID detector. The selected values are as follows: carrier gas was He (1.0 mL/min), temperature linearly increased from 40–315 °C (10 °C/min), injection volume: 1 μL, temperature: 250 °C, temperature (FID detector): 300 °C. CI was used as the ion source, with isobutane as the reagent gas. The mass spectrum was collected after a 4 min solvent delay.

## 5. Conclusions

*Pf*DyP B2 is an enzyme which can be easily expressed and used in peroxidase-based applications, ranging from assays and biosensors to dye-contaminated wastewater treatments. Notably, the first DyP-catalyzed insertion of a carbene into an N-H bond was reported. Although this is a preliminary result it could be a good platform for future applications of *Pf*DyP B2 in the synthesis of valuable compounds.

**Supplementary Materials:** The following are available online at http://www.mdpi.com/2073-4344/9/5/463/s1, GC-MS spectra of the synthesized compound.

**Author Contributions:** Conceptualization, N.L. and N.B.; methodology, N.L.; investigation, N.L., N.D., E.R., S.S., I.O., and N.B.; resources, M.F. and Z.V.; writing—original draft preparation, N.L.; writing—review and editing, all authors.

**Funding:** This research was funded by the Serbian Ministry of Education, Science and Technological Development, grant number ON172048.

**Conflicts of Interest:** The authors declare no conflict of interest.

## References


1. Fraaije, M.W.; Bloois, E.V. DyP-type peroxidases: A promising and versatile class of enzymes. *Enzym. Eng.* **2012**, *1*, 1–3. [CrossRef]
2. Colpa, D.I.; Fraaije, M.W.; van Bloois, E. DyP-type peroxidases: A promising and versatile class of enzymes. *J. Ind. Microbiol. Biotechnol.* **2014**, *41*, 1–7. [CrossRef] [PubMed]
3. Lončar, N.; Colpa, D.I.; Fraaije, M.W. Exploring the biocatalytic potential of a DyP-type peroxidase by profiling the substrate acceptance of *Thermobifida fusca* DyP peroxidase. *Tetrahedron* **2016**, *72*, 7276–7281. [CrossRef]
4. Liers, C.; Bobeth, C.; Pecyna, M.; Ullrich, R.; Hofrichter, M. DyP-like peroxidases of the jelly fungus *Auricularia auricula-judae* oxidize nonphenolic lignin model compounds and high-redox potential dyes. *Appl. Microbiol. Biotechnol.* **2010**, *85*, 1869–1879. [CrossRef]
5. Linde, D.; Ruiz-Duenas, F.J.; Fernandez-Fueyo, E.; Guallar, V.; Hammel, K.E.; Pogni, R.; Martinez, A.T. Basidiomycete DyPs: Genomic diversity, structural-functional aspects, reaction mechanism and environmental significance. *Arch. Biochem. Biophys.* **2015**, *574*, 66–74. [CrossRef]
6. Salvachua, D.; Prieto, A.; Martinez, A.T.; Martinez, M.J. Characterization of a novel dye-decolorizing peroxidase (DyP)-type enzyme from *Irpex lacteus* and its application in enzymatic hydrolysis of wheat straw. *Appl. Environ. Microbiol.* **2013**, *79*, 4316–4324. [CrossRef] [PubMed]
7. Scheibner, M.; Hulsdau, B.; Zelena, K.; Nimtz, M.; de Boer, L.; Berger, R.G.; Zorn, H. Novel peroxidases of *Marasmius scorodonius* degrade beta-carotene. *Appl. Microbiol. Biotechnol.* **2008**, *77*, 1241–1250. [CrossRef]
8. van Bloois, E.; Torres Pazmino, D.E.; Winter, R.T.; Fraaije, M.W. A robust and extracellular heme-containing peroxidase from *Thermobifida fusca* as prototype of a bacterial peroxidase superfamily. *Appl. Microbiol. Biotechnol.* **2010**, *86*, 1419–1430. [CrossRef]
9. Habib, M.H.; Rozeboom, H.J.; Fraaije, M.W. Characterization of a new DyP-Peroxidase from the alkaliphilic cellulomonad, *Cellulomonas bogoriensis*. *Molecules* **2019**, 24. [CrossRef]
10. Colpa, D.I.; Lončar, N.; Schmidt, M.; Fraaije, M.W. Creating Oxidase-Peroxidase Fusion Enzymes as a Toolbox for Cascade Reactions. *ChemBioChem* **2017**, *18*, 2226–2230. [CrossRef]
11. Habib, M.H.M.; Deuss, P.J.; Lončar, N.; Trajkovic, M.; Fraaije, M.W. A biocatalytic one-pot approach for the preparation of lignin oligomers using an oxidase/peroxidase cascade enzyme system. *Adv. Synth. Catal.* **2017**, *359*, 3354–3361. [CrossRef]
12. Yu, W.; Liu, W.; Huang, H.; Zheng, F.; Wang, X.; Wu, Y.; Li, K.; Xie, X.; Jin, Y. Application of a novel alkali-tolerant thermostable DyP-type peroxidase from *Saccharomonospora viridis* DSM 43017 in biobleaching of eucalyptus kraft pulp. *PLoS ONE* **2014**, *9*, e110319. [CrossRef] [PubMed]
13. Kong, L.; Guo, D.; Zhou, S.; Yu, X.; Hou, G.; Li, R.; Zhao, B. Cloning and expression of a toxin gene from *Pseudomonas fluorescens* GcM5-1A. *Arch. Microbiol.* **2010**, *192*, 585–593. [CrossRef] [PubMed]
14. Coelho, P.S.; Brustad, E.M.; Kannan, A.; Arnold, F.H. Olefin cyclopropanation via carbene transfer catalyzed by engineered cytochrome P450 enzymes. *Science* **2013**, *339*, 307–310. [CrossRef] [PubMed]
15. Wang, Z.J.; Peck, N.E.; Renata, H.; Arnold, F.H. Cytochrome P450-catalyzed insertion of carbenoids into N-H bonds. *Chem. Sci.* **2014**, *5*, 598–601. [CrossRef] [PubMed]
16. Tyagi, V.; Fasan, R. Myoglobin-Catalyzed Olefination of Aldehydes. *Angew. Chem. Int. Ed.* **2016**, *55*, 2512–2516. [CrossRef] [PubMed]
17. Sreenilayam, G.; Fasan, R. Myoglobin-catalyzed intermolecular carbene N-H insertion with arylamine substrates. *Chem. Commun.* **2015**, *51*, 1532–1534. [CrossRef]
18. Weissenborn, M.J.; Löw, S.A.; Borlinghaus, N.; Kuhn, M.; Kummer, S.; Rami, F.; Plietker, B.; Hauer, B. Enzyme-catalyzed carbonyl olefination by the *E. coli* protein YfeX in the absence of phosphines. *ChemCatChem* **2016**, *8*, 1636–1640. [CrossRef]
19. Letoffe, S.; Heuck, G.; Delepelaire, P.; Lange, N.; Wandersman, C. Bacteria capture iron from heme by keeping tetrapyrrol skeleton intact. *Proc. Natl. Acad. Sci. USA* **2009**, *106*, 11719–11724. [CrossRef]
20. Dailey, H.A.; Septer, A.N.; Daugherty, L.; Thames, D.; Gerdes, S.; Stabb, E.V.; Dunn, A.K.; Dailey, T.A.; Phillips, J.D. The *Escherichia coli* protein YfeX functions as a porphyrinogen oxidase, not a heme dechelatase. *MBio* **2011**, *2*, e00248-e11. [CrossRef]

21. Liu, X.; Yuan, Z.; Wang, J.; Cui, Y.; Liu, S.; Ma, Y.; Gu, L.; Xu, S. Crystal structure and biochemical features of dye-decolorizing peroxidase YfeX from *Escherichia coli* O157 Asp(143) and Arg(232) play divergent roles toward different substrates. *Biochem. Biophys. Res. Commun.* **2017**, *484*, 40–44. [CrossRef]
22. Hock, K.J.; Knorrscheidt, A.; Hommelsheim, R.; Ho, J.; Weissenborn, M.J.; Koenigs, R.M. Tryptamine synthesis by iron porphyrin catalyzed C-H functionalization of indoles with diazoacetonitrile. *Angew. Chem. Int. Ed.* **2019**, *58*, 3630–3634. [CrossRef]
23. Santos, A.; Mendes, S.; Brissos, V.; Martins, L.O. New dye-decolorizing peroxidases from Bacillus subtilis and *Pseudomonas putida* MET94: Towards biotechnological applications. *Appl. Microbiol. Biotechnol.* **2014**, *98*, 2053–2065. [CrossRef]
24. Yoshida, T.; Sugano, Y. A structural and functional perspective of DyP-type peroxidase family. *Arch. Biochem. Biophys.* **2015**, *574*, 49–55. [CrossRef]
25. Sugano, Y.; Muramatsu, R.; Ichiyanagi, A.; Sato, T.; Shoda, M. DyP, a unique dye-decolorizing peroxidase, represents a novel heme peroxidase family: ASP171 replaces the distal histidine of classical peroxidases. *J. Biol. Chem.* **2007**, *282*, 36652–36658. [CrossRef]
26. Brown, M.E.; Barros, T.; Chang, M.C.Y. Identification and characterization of a multifunctional dye peroxidase from a lignin-reactive bacterium. *ACS Chem. Biol.* **2012**, *7*, 2074–2081. [CrossRef]
27. Ahmad, M.; Roberts, J.N.; Hardiman, E.M.; Singh, R.; Eltis, L.D.; Bugg, T.D. Identification of DypB from *Rhodococcus jostii* RHA1 as a lignin peroxidase. *Biochemistry* **2011**, *50*, 5096–5107. [CrossRef]
28. Rahmanpour, R.; Bugg, T.D. Characterisation of Dyp-type peroxidases from *Pseudomonas fluorescens* Pf-5: Oxidation of Mn(II) and polymeric lignin by Dyp1B. *Arch. Biochem. Biophys.* **2015**, *574*, 93–98. [CrossRef]
29. Pesic, M.; Lopez, C.; Lopez-Santin, J.; Alvaro, G. From amino alcohol to aminopolyol: One-pot multienzyme oxidation and aldol addition. *Appl. Microbiol. Biotechnol.* **2013**, *97*, 7173–7183. [CrossRef]
30. de Gonzalo, G.; Colpa, D.I.; Habib, M.H.; Fraaije, M.W. Bacterial enzymes involved in lignin degradation. *J. Biotechnol.* **2016**, *236*, 110–119. [CrossRef]
31. Yang, C.; Yue, F.; Cui, Y.; Xu, Y.; Shan, Y.; Liu, B.; Zhou, Y.; Lu, X. Biodegradation of lignin by *Pseudomonas* sp. Q18 and the characterization of a novel bacterial DyP-type peroxidase. *J. Ind. Microbiol. Biotechnol.* **2018**, *45*, 913–927. [CrossRef] [PubMed]
32. Habib, M.; Trajkovic, M.; Fraaije, M.W. The biocatalytic synthesis of syringaresinol from 2,6-dimethoxy-4-allylphenol in one-pot using a tailored oxidase/peroxidase system. *ACS Catal.* **2018**, *8*, 5549–5552. [CrossRef] [PubMed]
33. Wasak, A.; Drozd, R.; Struk, Ł.; Grygorcewicz, B. Entrapment of DyP-type peroxidase from *Pseudomonas fluorescens* Pf-5 into Ca-alginate magnetic beads. *Biotechnol. Appl. Biochem.* **2018**, *65*, 238–245. [CrossRef] [PubMed]

*catalysts*

MDPI

*Article*

# Identification and Characterization of New Laccase Biocatalysts from *Pseudomonas* Species Suitable for Degradation of Synthetic Textile Dyes

**Mina Mandic [1], Lidija Djokic [1], Efstratios Nikolaivits [2], Radivoje Prodanovic [3], Kevin O'Connor [4], Sanja Jeremic [1,*], Evangelos Topakas [2,5,*] and Jasmina Nikodinovic-Runic [1,*]**

1 Institute of Molecular Genetics and Genetic Engineering, University of Belgrade, Vojvode Stepe 444a, 11000 Belgrade, Serbia
2 School of Chemical Engineering, National Technical University of Athens, 9 Iroon Polytechniou Str, Zoografou Campus, 15780 Athens, Greece
3 Faculty of Chemistry, University of Belgrade, Studentski Trg 12-16, 11000 Belgrade, Serbia
4 BEACON SFI Bioeconomy Research Centre and School of Biomolecular and Biomedical Science, University College Dublin, Belfield, Dublin 4, D4 Dublin, Ireland
5 Biochemical and Chemical Process Engineering, Division of Sustainable Process Engineering, Department of Civil, Environmental and Natural Resources Engineering, Luleå University of Technology, SE-97187 Luleå, Sweden
* Correspondence: sanjajeremic@imgge.bg.ac.rs (S.J.); vtopakas@chemeng.ntua.gr (E.T.); jasmina.nikodinovic@gmail.com or jasmina.nikodinovic@imgge.bg.ac.rs (J.N.-R.)

Received: 3 July 2019; Accepted: 17 July 2019; Published: 23 July 2019

check for updates

**Abstract:** Laccases are multicopper-oxidases with variety of biotechnological applications. While predominantly used, fungal laccases have limitations such as narrow pH and temperature range and their production via heterologous protein expression is more complex due to posttranslational modifications. In comparison, bacterial enzymes, including laccases, usually possess higher thermal and pH stability, and are more suitable for expression and genetic manipulations in bacterial expression hosts. Therefore, the aim of this study was to identify, recombinantly express, and characterize novel laccases from *Pseudomonas* spp. A combination of approaches including DNA sequence analysis, N-terminal protein sequencing, and genome sequencing data analysis for laccase amplification, cloning, and overexpression have been used. Four active recombinant laccases were obtained, one each from *P. putida* KT2440 and *P. putida* CA-3, and two from *P. putida* F6. The new laccases exhibited broad temperature and pH range and high thermal stability, as well as the potential to degrade selection of synthetic textile dyes. The best performing laccase was CopA from *P. putida* F6 which degraded five out of seven tested dyes, including Amido Black 10B, Brom Cresol Purple, Evans Blue, Reactive Black 5, and Remazol Brilliant Blue. This work highlighted species of *Pseudomonas* genus as still being good sources of biocatalytically relevant enzymes.

**Keywords:** laccase; genome-mining; heterologous expression; biocatalysis; *Pseudomonas*

## 1. Introduction

Laccases are blue multicopper enzymes (EC 1.10.3.2) that oxidize a broad range of both phenolic and non-phenolic substrates, via a four-electron reduction of molecular oxygen to water [1,2]. These enzymes are widely distributed in nature and have been isolated from bacteria [3–5], fungi [6,7] and plants [8,9], as well as from lichens [10], and sponges [11]. Although laccases are heterogeneous in different species, implying diversity in their function, four copper-binding motifs are conserved in the most laccases, especially bacterial forms [12]. The presence of different metal ions can affect laccase activity, either inducing or suppressing it. Metal ions such as $Cu^{+2}$, $Ca^{+2}$, $Ni^{+2}$, $Co^{+2}$, and $Mn^{+2}$ are generally known to accelerate laccase activity at a remarkable level [13].

Due to broad substrate spectrum, laccases have become very attractive for a variety of biotechnological and industrial applications such as organic synthesis; lignin degradation; and bio-product formation for the food, textile, and pharmaceutical industries; remediation of contaminated environments; as well as construction of biosensors and biofuel cells [14–16]. During the last decade laccases have been used in decolorization and detoxification of textile effluents [17]. Effluents from the textile industry are usually complex, containing a wide variety of synthetic dyes [18], among which the most common are azo dyes, anthraquinone, triphenylmethane, and indigo dyes [19].

In the recent years, development of high-performance recombinant bacterial strains and the possibility of increasing the production of recombinant proteins created new opportunities for the commercial use of laccases, since the production from wild type strains has limitations in growth and product yield, which are not suitable for standard industrial fermentations [20]. The majority of studies on laccases are focused on laccases originated from fungi. Having in mind that industrial processes often require high temperature and pressure, or extremely acidic or alkaline pH, fungal laccases are not the best candidates for such industrial applications since they operate in a temperature range from 30 °C to 55 °C and a slightly acidic pH range [2,21]. In addition, their heterologous expression is hindered by post-translational modifications [22], making their production cost-ineffective. Consequently, bacterial laccases are increasingly being sought for use in the industry due to the advantage of higher growth rates and better suitability for improvement of enzyme activity and expression level [23,24].

A number of bacterial laccases have been identified, heterologously expressed and studied at the molecular level. The first and the most studied bacterial laccase is CotA from *Bacillus subtilis* [25], followed by laccases from *B. coagulans*, *B. clausii* [26], and *B. licheniformis* [3]. The other significant group of bacterial laccases are from *Streptomyces* species, e.g., *S. coelicolor* [27], *S. cyaneus* [28], *S. bikiniensis* [29], and *S. ipomoea* [30].

Laccases from *Pseudomonas* species are mostly identified and purified from wild type producer strains [31–33] and until now, only one laccase from *P. putida* KT2440 was heterologously expressed and characterized [5]. Having in mind the importance of *Pseudomonas* strains, as a biotechnological platform for various industrial applications [34,35], as well as wealth and applicability of other enzymes from *Pseudomonas* species in biocatalysis, we set out to identify, recombinantly express, and characterize novel laccases from a number of *Pseudomonas* species, and assess their ability to degrade synthetic dyes widely used in textile industry.

## 2. Results and Discussion

### 2.1. Screening for Laccase Activity Using Guaiacol Agar Plates and ABTS Assay

Seven different *Pseudomonas* strains (*P. putida* F6, *P. putida* KT2440, *P. putida* CA-3, *P. putida* mt-2, *P. putida* S12, *P. chlororaphis* B561 and *P. aeruginosa* PAO1) from our laboratory collection were tested for laccase-like activity on agar plates containing standard laccase substrates guaiacol or syringaldazine 0.01% (*w/v*) and the development of dark colors due to the oxidation of these substrates was monitored [36]. In addition, supernatants and cell-free extracts of these strains previously grown in mineral MSM liquid medium supplemented with 5 mM phenylacetic acid to induce expression of enzymes from the aerobic catabolism of aromatic compounds [37] were tested in the ABTS assay (Section 3.4). Significantly higher enzyme activities were detected in the cell-free extracts than in supernatants of all tested *Pseudomonas* sp. cultures (data not shown), indicating the intracellular location of these enzymes, which is in line with previous literature reports on intracellular bacterial laccases [38,39].

Based on the fast reddish-brown color formation, indicating oxidation of guaiacol (Figure 1A), as well as the best activity of cell-free extracts in ABTS assay (Figure 1B), three strains, namely *P. putida*

F6, *P. putida* KT2440, and *P. putida* CA-3, were chosen for further study. Notably, cell-free extracts of *P. putida* F6 exhibited two times higher laccase activity (ABTS assay) compared to cell free extracts from *P. putida* KT2440 and *P. putida* CA-3, respectively, which showed a similar activity level (Figure 1B).

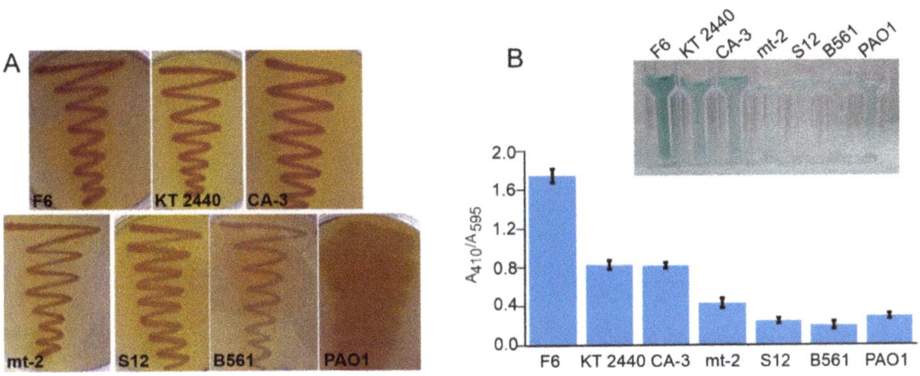


**Figure 1.** Laccase activity assay on guaiacol-containing plates (**A**) and in ABTS assay (**B**). Abbreviations F6, KT2440, CA-3, mt-2, S12, B561, and PAO1 refer to wild type strains *P. putida* F6, *P. putida* KT2440, *P. putida* CA-3, *P. putida* mt-2, *P. putida* S12, *P. chlororaphis* B561 and *P. aeruginosa* PAO1 (**A**) and appropriate cell-free extracts (**B**). Enzyme activity in ABTS assay was expressed as a ratio of product formation to the total protein concentration ($A_{410}/A_{595}$). Inlet in (**B**) represents color change in ABTS assay due to laccase activity from cell-free extracts.

## 2.2. Primer Design Based on UniProt Data Analysis

Based on data from UniProt database 15 laccase and laccase-like multicopper oxidase (MCO) protein sequences from *Pseudomonas* strains were aligned using ClustalΩ tool (https://www.ebi.ac.uk/Tools/msa/clustalo/) and degenerate primers were designed (Table 1). Genes coding for enzymes with laccase activity from *P. putida* KT2440 and *P. putida* CA-3 were successfully amplified using designed primers and identified as MCO by sequencing. On the other side, amplification products obtained for *P. putida* F6 were also sequenced, but did not show any similarity with any known laccases (data not shown). Based on the obtained sequences for laccase genes from *P. putida* KT2440 and *P. putida* CA-3, specific primers for amplification and cloning were designed (Table 1). Amplicons obtained from *P. putida* KT2440 and *P. putida* CA-3 with specific primers were sequenced and aligned (https://www.ebi.ac.uk/Tools/psa/emboss_needle/nucleotide) [40], and showed 96.8% identity. Given that presence of laccase-like protein in *P. putida* F6 has been indicated previously in the literature [41] and that *P. putida* F6 in this study showed at least two-times higher activity in ABTS assay (Figure 1B), further pursuit for this particular enzyme was undertaken, involving protein purification from the wild type strain, followed by protein and genome sequencing.

## 2.3. Purification of Protein with Laccase Activity from P. putida F6

Proteins from *P. putida* F6 with laccase activity were purified using ion-exchange chromatography. Fractions obtained were tested for enzyme activity towards ABTS in the presence of $CuSO_4$ (2 mM) with activity detected in two fractions (Figure 2). Active fractions were pooled and named Lacc1 and Lacc2, implying that they exhibit a laccase-like activity and their activity was improved drastically in the presence of $CuSO_4$. Both fractions were additionally purified, concentrated, and checked for homogeneity by SDS-PAGE. Lacc1 showed one homogenous band on ~24 kDa (Figure 2A), while the lacc2 fraction was non-homogeneous and lost most of its activity after two gel filtration steps (Figure 2B). Therefore, the Lacc1 fraction was used for further studies.

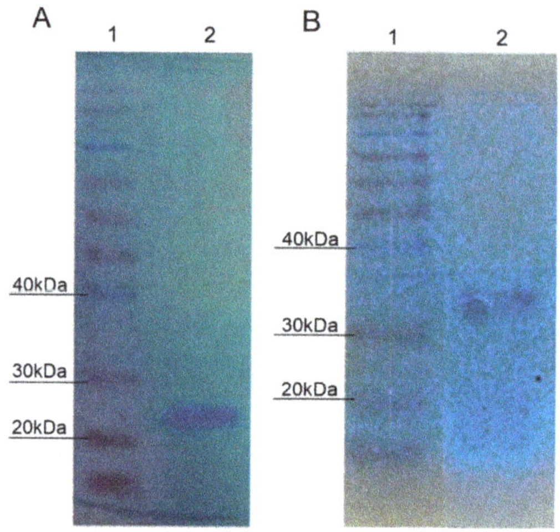


**Figure 2.** SDS-PAGE of partially purified (enriched fractions) enzymes from *P. putida* F6. (**A**) lane 1—protein standard, lane 2—Lacc1 fraction (~24 kDa); (**B**) lane 1—protein standard, lane 2—Lacc2 fraction.

### 2.4. N-Terminal Lacc1 Protein and P. putida F6 Genome Sequencing

Lacc1 fraction from *P. putida* F6 was used for N-terminal sequencing that revealed MTHHSED motif on the N terminus. In addition, the whole genome of *P. putida* F6 was sequenced and annotated using the PROKKA software tool [42]. Analysis of the whole genome by a sequence-based search for proteins with laccase-like activity and a structure-based search for conserved protein domains using Conserved Domains tool (http://www.ncbi.nlm.nih.gov/Structure//cdd/wrpsb.cgi) [43,44] revealed eight different laccase-like gene sequences (Table 2). Amongst them, only one contained MTHHSED motif, namely gene coding for a four-helix bundle copper-binding protein (*cbp*). The Conserved Domains tool detected multicopper oxidase (*copA*) which was also of high interest due to the fact that both purified fractions showed a certain size and $CuSO_4$ dependant activity (Figure 2). Based on the *copA* and *cbp* sequences, specific primers for cloning and heterologous expression were designed (Table 1). It is worth mentioning that partially purified Lacc1 protein fraction (Figure 2A) from wild type *P. putida* F6 has a molecular weight of approximately 24 kDa judging from the SDS-PAGE analysis (Figure 2A). The identified *cbp* gene was 348 bp in length, which predictably corresponds to approximately 12.7 kDa. This gene is followed by *copC* gene that is 381 bp in length, which predicts to a protein of approximately 13.9 kDa. The close proximity of these two genes (separated by 11 bp) could result in possible transcription of both genes (*cbp* and *copC*) and translation to a single peptide which in turn may explain the discrepancy in predicted size of Cbp protein (Table 2) and Lacc1 protein fraction (Figure 2A).

**Table 1.** Oligonucleotide sequences and the annealing temperaturesused for the amplification of laccases from *Pseudomonas* species.

| | Primers with Restriction Enzyme Sites | Annealing T | Strain |
|---|---|---|---|
| PS_LACF<br>PS_LACR | ATGAGTGRCCTGRCBCAG<br>GCGGNTCCAGCCASACCARSGA | 50 °C | *Pseudomonas* spp. |
| CA3F<br>CA3R | TAACA**GGATCC**GAGTGGCCTGACTCAGG (Bam HI)<br>TAATT**AAGCTT**TGCGGTTCCAGCCAGAC (Hind III) | 55 °C | *P. putida* CA-3 |
| KTF<br>KTR | TAACA**GGATCC**GAGTGACCTGACGCAGG (Bam HI)<br>TAATT**AAGCTT**GCGCGGGTCCAGCCAGAC (Hind III) | 59 °C | *P. putida* KT2440 |
| CopAF<br>CopAR | TAACA**GCTAGC**ATGTCGCATGATGATTTTCGT (Nhe I)<br>TAATT**AAGCTT**TTCGTCGACCCTCACCGTGCG (Hind III) | 55 °C | *P. putida* F6 |
| CbpF<br>CbpR | TAACA**GCTAGC**ATGACTCACCATTCCGAAGAC (Nhe I)<br>TAATT**AAGCTT**AGCCGCCATGGCGCTGCAGCT (Hind III) | 58 °C | *P. putida* F6 |

**Table 2.** Genes coding for proteins with laccase-like activity discovered in genome of *P. putida* F6.

| Gene Name | Protein Size (AA) | GeneBank Match | GeneBank Acc No | Identity (%) | Discovered with |
|---|---|---|---|---|---|
| Putative cysteine-rich protein YhjQ | 116 | Four-helix bundle copper-binding protein (Cbp) from *Pseudomonas putida* | WP_026070601 | 100 | MTHHSED motif search |
| Not annotated | 623 | Copper resistance system multicopper oxidase (CopA) from *Xantomonadaceae* | WP_017354985 | 100 | Conserved Domains tool |
| Copper resistance protein B | 425 | Copper resistance protein B (CopB) from *Xantomonadaceae* | WP_017354984 | 100 | Genome annotation |
| Copper resistance protein C | 127 | Copper homeostasis periplasmic binding protein (CopC) from *Xanthomonadaceae* | WP_017354979 | 100 | Genome annotation |
| Laccase domain protein YfiH | 246 | Multi-copper polyphenol oxidoreductase laccase from *Pseudomonas putida* | WP_075804457 | 96.3 | Genome annotation |
| Laccase domain protein YfiH | 256 | Multi-copper polyphenol oxidoreductase laccase from *Stenotrophomonas maltophilia* | WP_099560244 | 99.2 | Genome annotation |
| Multicopper oxidase mco | 460 | Multicopper oxidase from *Pseudomonas putida* | WP_075806698 | 97.4 | Genome annotation |
| Blue copper oxidase CueO | 675 | Multicopper oxidase protein from *Pseudomonas* | WP_075804455 | 79.9 | Genome annotation |

### 2.5. Cloning and Expression of Enzymes with Laccase Activity from Pseudomonas Species

Based on the sequence analysis and activity data, two amplicons from *P. putida* F6 (*cbp*—348 bp and *copA*—1869 bp) and amplicons from *P. putida* KT2440 (*mcoKT*—741 bp) and *P. putida* CA-3 (*mcoCA3*—738 bp) were cloned into pRSET B expression vector. Sequences of four amplicons are deposited in GenBank under accession numbers MN075141, MN075142, MN075139, and MN075140. The recombinant proteins were expressed as N-terminal $His_6$ fusion proteins in *E. coli* Rosetta (DE3) after 72 h induction at 17 °C with 1 mM IPTG. Activity of overexpressed proteins was determined in ABTS assay using cell-free extracts. Proteins of interest were purified with Ni-NTA affinity chromatography, and their activity was confirmed using zymography assay (Figure 3).

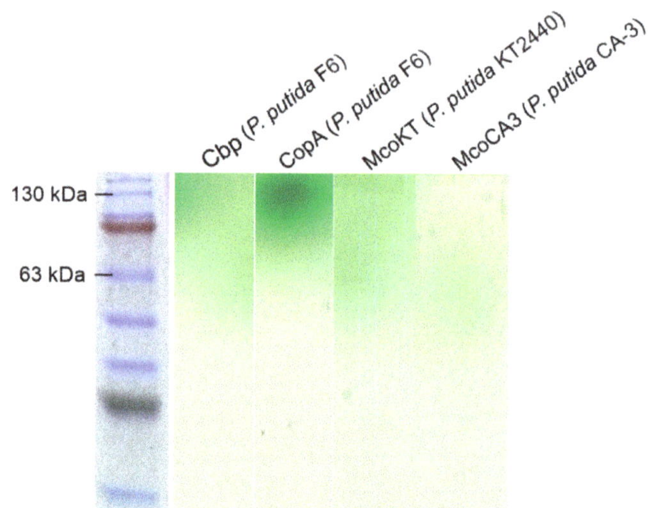


**Figure 3.** Zymography assay of purified laccases from *Pseudomonas* species using native PAGE analysis. The first lane—BlueStar prestained protein marker (Nippon genetics). Lanes from left to right: Cbp (*P. putida* F6), CopA (*P. putida* F6), McoKT (*P. putida* KT2440), and McoCA3 (*P. putida* CA-3).

Although most of the laccases studied thus far are of fungal origin, they have limitations including difficulties in heterologous protein expression due to the need for posttranslational modifications, as well as narrow temperature and pH range [4,14]. During the last decade bacterial laccases gained more attention [31,45] due to the number of advantages over fungal laccases, one of which is being more amenable for recombinant expression and directed evolution studies in an *E. coli* host. Despite wide use of *Pseudomonas* enzymes in different industrial applications, to date only one laccase-like enzyme, namely CopA from *P. putida* KT2440, was recombinantly expressed and characterized [5]. The 247 amino acid multicopper oxidase enzyme identified from *P. putida* KT2440 in the current study is much smaller than the 669 amino acid enzyme identified by Granja-Travez and co-workers [5].

### 2.6. Characterization of Recombinantly Expressed Laccases from Pseudomonas species

Activity profiles of four purified laccases on different pH values were determined using ABTS as the substrate and buffers of different pHs ranging from 3 to 10. All tested enzymes exhibited a broad pH range, pH 3 to pH 8 for Cbp and CopA, and pH 3 to pH 7 for McoKT and McoCA3, all showing maximum activity at pH 4 (Figure 4A). Obtained results are in accordance with available literature data: CopA from *P. putida* KT2440, laccase from *Bacillus coagulans*, and *B. clausii* laccase [26] exhibited maximum activity on ABTS as a substrate on pH 4 [5]. Furthermore, laccase from *B. licheniformis* showed maximal activity at pH 3 [3] while laccase from *Streptomyces cyaneus* had the maximum activity at pH 5 [4].

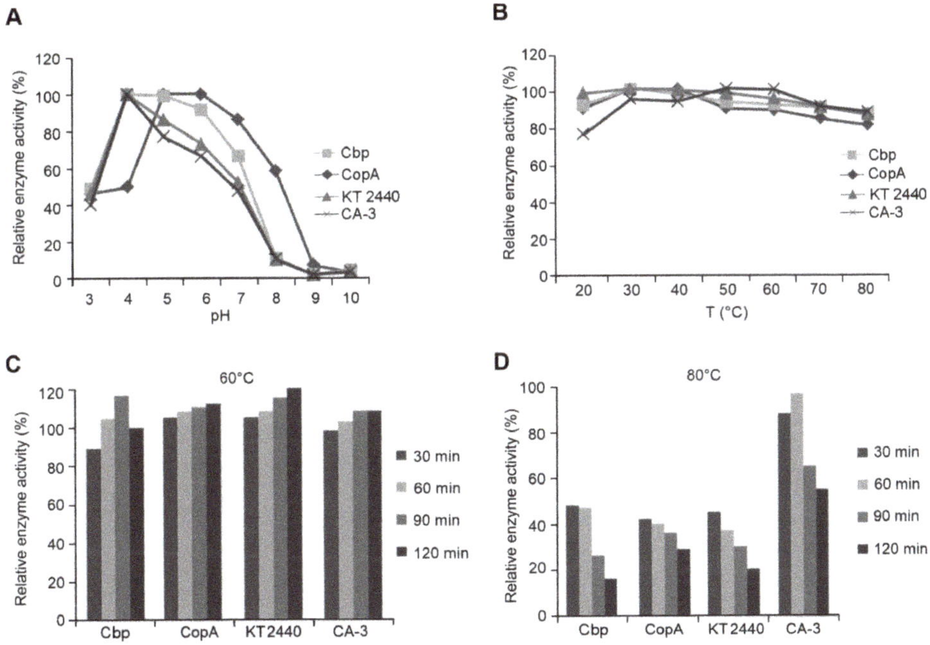


**Figure 4.** pH, temperature optimum and temperature stability of expressed laccases from *Pseudomonas* species: (**A**) relative enzymes activity at different pH values, (**B**) relative enzymes activity at different temperatures, (**C**) temperature stability at 60 °C, (**D**) temperature stability at 80 °C.

Temperature optimum was determined by incubating the enzymes at temperatures ranging from 20 to 80 °C under previously determined optimal pH. All tested laccases showed a broad temperature range from 20 to 80 °C with optimal range from 30 to 50 °C (Figure 4B). High enzyme activity even at elevated temperatures correlates with other bacterial laccases and laccase-like multi-copper oxidases [26,27].

Temperature stability of purified laccases from *Pseudomonas* species was assessed by incubation of enzymes at 60 °C and 80 °C for a total of 120 min, while taking enzyme aliquots for activity tests with ABTS every 30 min. All tested enzymes exhibited high temperature stability at 60 °C, since their activities were slightly elevated in comparison to control (enzyme activity measured before incubation at 60 °C and 80 °C) that was arbitrarily set to 100%, even after 2h of incubation (Figure 4C). Temperature activation of enzyme is also previously shown by Ece and coworkers [4]. McoCA3 enzyme exhibited the best temperature stability overall, as it retained 55% of its initial activity after 120 min at 80 °C, while Cbp, CopA, and McoKT retained 16%, 29%, and 20% of activity, respectively (Figure 4D).

Considering that *Pseudomonas* strains are mesophilic microorganisms, enzymes tested in this study showed high thermotolerance, which is in accordance with literature data on other enzymes from mesophilic microorganisms. Lončar and coworkers tested the activity of purified recombinant laccase from *Bacillus licheniformis* and retained 50% of activity after 100 min incubation at 60 °C, and 60 min incubation at 70 °C [3]. Cell-free extracts of heterologously expressed laccase-like MCO from *Bacillus* strains retained their activities after 30 min of incubation at 70 °C [26]. Ece and coworkers tested the activity of purified recombinant laccase from *Streptomyces* strain on DMP (2,6-dimethoxyphenol) substrate and showed retention of 50% of activity after 1 h incubation at 60 °C, while almost all activity was lost at 90 °C [4].

## 2.7. Degradation of Synthetic Textile Dyes Using Laccases from Pseudomonas Species

Potential of recombinant laccases from *Pseudomonas* strains to degrade seven synthetic dyes was followed spectrophotometrically for four days and results are presented as relative dye degradation in percents (Figure 5, only degraded dyes are shown). All dyes selected for degradation experiments are widely used in textile and dyeing industries. Among the seven tested dyes, five belong to azo dyes, which make up to 70% of all textile dyestuff produced [46]. Although, degradation experiments were first performed without the addition of a redox mediator (data not shown), the addition of ABTS (0.05 mM) substantially improved degradation of synthetic dyes by tested laccases. Two dyes, namely Erythrosin B and Orange G, were only dyes that were not degraded by tested laccases from *Pseudomonas* species. Among tested samples the best performing was cell-free extract containing CopA laccase from *P. putida* F6. CopA laccase degraded five out of seven tested dyes, namely Amido Black 10B, Brom Cresol Purple, Evans Blue, Reactive Black 5, and Remazol Brilliant Blue (Figure 5B). This laccase completely degraded azo dye Reactive Black 5 within four days, while 86% of dye degradation was detected after 48 h (Figure 5B). Pereira and coworkers [47] tested the ability of purified recombinant bacterial CotA-laccase from *Bacillus subtilis* to degrade Reactive Black 5 and obtained 85% of degradation after 24 h. Among seven tested dyes, Amido Black 5 was the only dye significantly degraded by all used laccases, with a degradation range from 70 to 95% within four days. Mishra and coworkers tested three *Bacillus* soil isolates and obtained 50–84% of azo dye Amido Black 5 degradation after four days incubation with bacterial culture [48]. Ninety-five percent of triphenylmethane dye Brom Cresol Purple was degraded within four days incubation with cell-free extract containing laccase from *P. putida* CA-3 (Figure 5D). The majority of studies on textile dyes degradation were performed using fungal laccases. All laccases tested in this study showed promising potential for the application in textile dye degradation, having in mind that presented results were obtained using cell-free extracts of recombinant strains expressing laccases, not purified enzymes to make the process scalable and to avoid costly enzyme purification procedures.

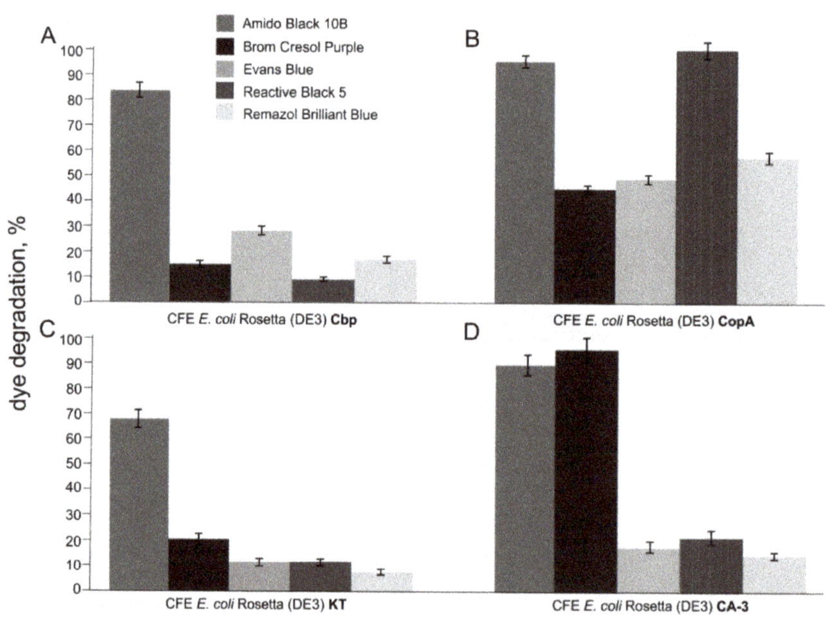


**Figure 5.** Degradation of five synthetic textile dyes using cell-free extracts (CFE) of *E. coli* Rosetta (DE3) expressing laccases: (**A**) Cbp from *P. putida* F6; (**B**) CopA from *P. putida* F6; (**C**) KT from *P. putida* KT2440; and (**D**) CA-3 from *P. putida* CA-3.

## 3. Materials and Methods

### 3.1. Reagents

All reagents and solvents used in this work were purchased either from Merck (Darmstadt, Germany) or Sigma-Aldrich (St. Louis, MO, USA). Dyes were purchased from Carl Roth (Karlsruhe, Germany), except Reactive Black 5 which was purchased from Sigma-Aldrich (St. Louis, MO, USA). GeneJet PCR purification, GeneJet gel extraction kit, and GeneJet plasmid purification kits were purchased from Thermo Fisher Scientific (Waltham, MA, USA).

### 3.2. Bacterial Strains

All strains used in this study as sources of laccases as well expression strains are presented in Table 3.

**Table 3.** Bacterial strains used in this study.

| Bacterial Strain | Reference |
|---|---|
| *Pseudomonas aeruginosa* PAO1 | ATCC 15,692 |
| *Pseudomonas putida* CA-3 | NCIMB 41,162 |
| *Pseudomonas putida* F6 | [33] |
| *Pseudomonas putida* KT2440 | ATCC 47,054 |
| *Pseudomonas putida* mt-2 | NCIMB 10,432 |
| *Pseudomonas putida* S12 | ATCC 70,0801 |
| *Pseudomonas chlororaphis* B561 | ATCC 19,523 |
| *Escherichia coli* Rosetta (DE3) | Merck, Darmstadt, Germany |

### 3.3. Media

Mineral Salts Medium (MSM) containing 9 g/L $Na_2HPO_4 \times 12H_2O$, 1.5 g/L $KH_2PO_4$, 0.2 g/L $MgSO_4 \times 7H_2O$, 0.002 g/L $CaCl_2$, 1 g/L $NH_4Cl$, 1 mL salt solution [49], supplemented with 0.7% casamino acids, 1% glucose, and 5mM phenylacetic acid was used for growth of wild type *Pseudomonas* strains and induction of laccases.

Luria–Bertani (LB) broth containing 10 g/L tryptone, 10 g/L NaCl, and 5 g/L yeast extract was used for the growth. Laccase activity was tested on agar plates containing 10 g/L tryptone, 10 g/L NaCl, 5 g/L yeast extract, and 15 g/L agar with 0.01% guaiacol and 0.35 mM $CuSO_4$. Plates were incubated for 7 days at 30 °C.

For selection of clones Luria Agar (LA) plates supplemented with 100 µg/mL of ampicillin. For expression of recombinant laccase terrific broth (TB) was used containing 24 g/L yeast extract, 20 g/L tryptone, 0.4% glycerol, 0.017 M $KH_2PO_4$, and 0.072M $K_2HPO_4$ with 100 µg/mL ampicillin.

### 3.4. Screen for Laccase Activity on Guaiacol Agar and in ABTS Assay

The plate method assay for laccase activity was performed on LA agar supplemented with 0.01%(*w/v*) guaiacol of syringaldazine as laccase substrates and 0.35 mM $CuSO_4$. Strains were seeded and grown overnight at 37 °C and observed for the formation of brown colored zones around colonies that form when laccase is produced by the tested strain.

2,2′-azino-bis(3-ethylbenzothiazoline-6-sulphonic acid) (ABTS) assay was performed either with supernatants and cell-free extracts of wild type strains or with heterologously expressed purified proteins. The reaction was performed in 100 mM Na-acetate buffer pH 4.5 with the addition of 1 mM ABTS and 2 mM $CuSO_4$. Absorption of formed product was followed spectrophotometrically at 410 nm.

### *3.5. Primer Design, Gene Amplification and Cloning*

Based on the data retrieved from UniProt database search for laccase from *Pseudomonas*, primers for laccase amplification from selected *Pseudomonas* strains were designed. All primers are presented in Table 1, where restriction sites for positional cloning in pRSETB vector are designated in bold.

PCR amplification of genes coding for proteins with laccase activity was carried out based on the manufacturer's instructions of GeneJet PCR kit (Thermo Fisher Scientific) on corresponding primer specific annealing temperature (Table 1). Amplified and restriction digested PCR fragments were positionally cloned in pRSET B vector and *E. coli* Rosetta (DE3) competent cells (both from Thermo Fisher Scientific) were transformed.

### *3.6. Protein Purification from P. putida F6*

A functional isolation of protein with laccase (ABTS oxidizing) activity was performed using anion-exchange (Q-Sepharose) columns were purchased from Amersham (Amersham Biosciences AB, Uppsala, Sweden). Cell-free extracts of phenylacetic acid-induced (5 mM) *P. putida* F6 cells were obtained using following procedure: cell culture was centrifuged for 20 min at 3000× *g* (Eppendorf centrifuge 5804, Hamburg, Germany). Pellet was resuspended in lysis buffer (20 mM Tris-HCl, pH 8, and 1 mg/mL lysozyme) and incubated 30 min at 37 °C. Mixture was sonicated at 20 KHz, 3 pulses of 15 s (Soniprep 150, MSE (UK) Ltd., Lindon, UK). Cell-free extracts were clarified by centrifugation for 40 min at 20,817× *g*, 4 °C (Eppendorf Centrifuge 5417 R, Hamburg, Germany). Obtained cell-free extracts were dialyzed against 20 mM Tris-HCl buffer pH 8 and loaded on a manually packed (5 cm × 25 cm) strong anion-exchange Q-Sepharose column pre-equilibrated at the same buffer. Elution was performed applying a linear gradient of 0–0.5 M NaCl at a flow rate of 5 mL/min. Partially purified enzymes were equilibrated in 20 mM Tris-HCl pH 7 buffer, applied on pre-equilibrated column separately and eluted as previously described. Concentrated fractions were checked for homogeneity by SDS-PAGE, stained with Coomassie Brilliant Blue. Partially purified enzymes were loaded onto a HiPrep 16/60 Sephacryl S-200 HR (1.6 cm × 60 cm) and HiPrep Sephacryl S-300 HR 26/60 (2.6 cm × 60 cm) gel filtration columns, previously equilibrated with 20 mM Tris-HCl pH 7 with 150 mM NaCl at a flow rate 60 mL/h. Obtained fractions were tested for enzyme activity using ABTS assay (Section 3.4).

### *3.7. N-Terminal and Genome Sequencing*

Purified laccase from *Pseudomonas putida* F6 was sequenced by Cambridge Peptides Ltd., West Midlands, UK. A sequence of 7 amino acids at N-terminus was identified by the Edman degradation method.

Genome of *Pseudomonas putida* F6 was sequenced in MicrobesNG company (Birmingham, UK). Obtained raw genome sequences were annotated using PROKKA software tool and analyzed using Conserved Domains tool [42–44]. Genes of interest were discovered by motif, sequence and domain-based search and primers for cloning in pRSETB were designed (Table 1).

### *3.8. Recombinant Protein Expression and Purification*

Host cells carrying pRETB plasmid with laccases from different *Pseudomonas* strains were grown overnight at 37 °C in TB medium containing 100 g/mL ampicillin. Overnight culture was diluted (1%) into TB medium containing ampicillin (100 g/mL) and $CuSO_4$ at final concentration of 2 mM and after reaching OD600 = 0.5 expression was induced with 1 mM IPTG and carried out for 48 h at 17 °C with shaking at 180 rpm. Cells were pelleted for 10 min at 3000× *g* (Eppendorf centrifuge 5804, Hamburg, Germany). Lysis buffer containing (50 mM $NaH_2PO_4$, pH 8, 300 mM NaCl, 10 mM imidazole, and 1 mg/mL lysozyme) was used for the preparation of cell-free extracts. The mixture was incubated for 30 min at 37 °C followed by sonication of 3 pulses of 15 s at 20 KHz (Soniprep 150, MSE (UK) Ltd., England). Cell-free extracts were clarified by centrifugation for 40 min at 20 817× *g*, 4 °C (Eppendorf Centrifuge 5417 R, Hamburg, Germany). Clear supernatant was loaded on a NiNTA agarose (Qiagen, Hilden, Germany) equilibrated with 50 mM $NaH_2PO_4$ pH8, 300 mM NaCl, 10 mM imidazole. Non-specifically bound proteins were washed with one volume of 50 mM imidazole in starting buffer followed by elution with 250 mM and 500 mM imidazole in the same buffer. Total protein concentration in collected fractions was determined according to the Bradford method [50] using Quick StartTM Bradford 1× Reagent (BioRad Laboratories, California, USA). Collected fractions were analyzed by activity assay on ABTS and SDS-PAGE.

### 3.9. Temperature and pH Optimum of Purified Laccases

The optimum pH of the laccase enzymes was determined within a pH range of 3 to 10 using ABTS (2,2′-azino-bis(3-ethylbenzothiazoline-6-sulphonic acid) as a substrate, and purified proteins with adjusted concentrations. For pH range from 3 to 5, acetate buffer was used; for pH 6 and 7, 20 mM phosphate buffer was used; while for pH 8 and pH 9 and 10, 20 mM Tris-HCl and 20 mM glycine-NaOH were used, respectively. Temperature optimum of enzyme activity was measured in the range of 20 to 80 °C at previously determined optimal pH and buffer and with ABTS as a substrate. ABTS assay contained 20 mM Na-acetate buffer pH 4.5, 100 µL of sample, 1 mM ABTS, and 2 mM $CuSO_4$ and after incubation of 15 min at 37 °C reactions were followed spectrophotometrically for product formation at 420 nm.

Thermal stability of purified laccase was determined by measuring residual activity upon incubating aliquots of purified laccases at 60 °C and 80 °C. Samples were withdrawn at 30, 60, 90, and 120 min, placed on ice, and enzyme activity was determined in ABTS assay as described in Section 3.4. For testing thermal stability 100 mM Na-acetate buffer of pH 4.0—pH 5.0 was used.

### 3.10. Textile Dyes Degradation

Seven different dyes were selected to study degradation ability of the laccases from *Pseudomonas* species. Selected dyes belong to three different groups: Amido Black 10B, Evans Blue, Reactive Black 5, Remazol Briliant Blue, and Orange G are azo dyes, Brom Cresol Purple is triphenylmethane dye, while Erythrosin B is xanthene dye. Stock solutions of the dyes in 100 mM Na-acetate buffer pH 4.5 were stored in the dark at 4 °C. For dyes degradation experiments cell-free extracts of four recombinant strains expressing laccases from *Pseudomonas* species were used. Reaction consisted of cell-free extracts of adjusted protein concentrations in 100 mM Na-acetate buffer pH 4.5, 0.05 mM ABTS, and 2 mM $CuSO_4$ and corresponding dye. Concentration of each dye was set to correspond to 0.6 absorbance units at the maximum wavelength of a specific dye. In this experiment two controls were set up. One control reaction consisted of appropriate dye, 100 mM Na-acetate buffer pH 4.5, 0.05 mM ABTS, and 2 mM $CuSO_4$, and was used to monitor dye degradation over time. Second control reaction, used as blank, was prepared for each laccase, and consisted of cell-free extract in 100 mM Na-acetate buffer pH 4.5, 0.05 mM ABTS, and 2 mM $CuSO_4$. Reactions were incubated for four days at 30 °C, and dye concentrations were spectrophotometrically measured at maximum wavelength determined for each dye. Dye degradation (%) was calculated using following equation:

$$\frac{initial\ absorbance - final\ absorbance}{initial\ absorbance} \times 100\% \quad (1)$$

## 4. Conclusions

Four active recombinant laccases from three *Pseudomonas* sp. (*P. putida* KT2440 and *P. putida* CA-3, and *P. putida* F6) were discovered using different molecular approaches including DNA sequence analysis, N-terminal protein sequencing, and genome sequencing data analysis. These new laccases exhibited good stability at elevated temperatures, and were active over a broad temperature and pH range. All tested laccases showed good potential in degradation of synthetic textile dyes, with CopA from *P. putida* F6 being the most promising, as this laccase was active on five out of seven tested dyes, namely Amido Black 10B, Brom Cresol Purple, Evans Blue, Reactive Black 5, and Remazol Brilliant

Blue. This research confirmed that *Pseudomonas* genus is still a source of biocatalytically relevant enzymes, since all tested enzymes showed promising potential for the degradation of textile dyes which makes them good candidates for the application in bioremediation.

**Author Contributions:** E.T., S.J., and J.N.-R. designed and supervised this study. M.M. performed experimental work. E.N, R.P., and K.O. were involved in protein purification and characterization from wild type strain. L.D. performed protein and genome data analysis. All authors contributed in the preparation of the manuscript.

**Funding:** This work was supported by the Ministry of Education, Science and Technological Development, Republic of Serbia, under Grant No. 173048. FEMS Research Grant (FEMS-RG-2016-0088) to Mina Mandic is acknowledged.

**Conflicts of Interest:** The authors declare no conflict of interest.

## References

1. Yaropolov, A.I.; Skorobogat'Ko, O.V.; Vartanov, S.S.; Varfolomeyev, S.D. Laccase: Properties, catalytic mechanism, and applicability. *Appl. Biochem. Biotechnol.* **1994**, *49*, 257–280. [CrossRef]
2. Morozova, O.V.; Shumakovich, G.P.; Gorbacheva, M.A.; Shleev, S.V.; Yaropolov, A.I. "Blue" laccases. *Biochemistry (Moscow)* **2007**, *72*, 1136–1150. [CrossRef] [PubMed]
3. Lončar, N.; Božić, N.; Vujičić, Z. Expression and characterization of a thermostable organic solvent-tolerant laccase from *Bacillus licheniformis* ATCC 9945a. *J. Mol. Catal. B Enzym.* **2016**, *134*, 390–395. [CrossRef]
4. Ece, S.; Lambertz, C.; Fischer, R.; Commandeur, U. Heterologous expression of a *Streptomyces cyaneus* laccase for biomass modification applications. *AMB Express* **2017**, *7*, 86. [CrossRef] [PubMed]
5. Granja-Travez, R.S.; Bugg, T.D.H. Characterisation of multicopper oxidase CopA from *Pseudomonas putida* KT2440 and *Pseudomonas fluorescens* Pf-5: Involvement in bacterial lignin oxidation. *Arch. Biochem. Biophys.* **2018**, *660*, 97–107. [CrossRef] [PubMed]
6. Brijwani, K.; Rigdon, A.; Vadlani, P.V. Fungal laccases: Production, function, and applications in food processing. *Enzyme. Res.* **2010**, *2010*, 149748–149758. [CrossRef] [PubMed]
7. Zerva, A.; Koutroufini, E.; Kostopoulou, I.; Detsi, A.; Topakas, E. A novel thermophilic laccase-like multicopper oxidase from *Thermothelomyces thermophila* and its application in the oxidative cyclization of 2′,3,4-trihydroxychalcone. *New Biotechnol.* **2019**, *49*, 10–18. [CrossRef]
8. Bryan, A.C.; Jawdy, S.; Gunter, L.; Gjersing, E.; Sykes, R.; Hinchee, M.A.; Winkeler, K.A.; Collins, C.M.; Engle, N.; Tschaplinski, T.J.; et al. Knockdown of a laccase in *Populus deltoides* confers altered cell wall chemistry and increased sugar release. *Plant Biotechnol. J.* **2016**, *14*, 2010–2020. [CrossRef]
9. Liu, Q.; Luo, L.; Wang, X.; Shen, Z.; Zheng, L. Comprehensive analysis of rice laccase gene (*OsLAC*) family and ectopic expression of *OsLAC10* enhances tolerance to copper stress in *Arabidopsis*. *Int. J. Mol. Sci.* **2017**, *18*, 209. [CrossRef]
10. Laufer, Z.; Beckett, R.P.; Minibayeva, F.V.; Lüthje, S.; Böttger, M. Diversity of laccases from lichens in suborder Peltigerineae. *Bryologist* **2009**, *112*, 418–426. [CrossRef]
11. Li, Q.; Wang, X.; Korzhev, M.; Schröder, H.C.; Link, T.; Tahir, M.N.; Diehl-Seifert, B.; Müller, W.E. Potential biological role of laccase from the sponge *Suberites domuncula* as an antibacterial defense component. *Biochim. Biophys. Acta Gen. Subj.* **2015**, *1850*, 118–128. [CrossRef] [PubMed]
12. Jin, L.; Yang, X.; Sheng, Y.; Cao, H.; Ni, A.; Zhang, Y. The second conserved motif in bacterial laccase regulates catalysis and robustness. *Appl. Microbiol. Biotechnol.* **2018**, *102*, 4039–4048. [CrossRef] [PubMed]
13. Muthukumarasamy, N.P.; Jackson, B.; Raj, A.J.; Sevanan, M. Production of extracellular laccase from *Bacillus subtilis* MTCC 2414 using agroresidues as a potential substrate. *Biochem. Res. Int.* **2015**, *2015*, 765190. [CrossRef] [PubMed]
14. Piscitelli, A.; Pezzella, C.; Giardina, P.; Faraco, V.; Giovanni, S. Heterologous laccase production and its role in industrial applications. *Bioengineered* **2010**, *1*, 252–262. [CrossRef] [PubMed]
15. Zouraris, D.; Zerva, A.; Topakas, E.; Karantonis, A. Kinetic and amperometric study of the MtPerII peroxidase isolated from the ascomycete fungus *Myceliophthora thermophila*. *Bioelectrochemistry* **2017**, *118*, 19–24. [CrossRef] [PubMed]
16. Romero-Guido, C.; Baez, A.; Torres, E. Dioxygen activation by laccases: Green chemistry for fine chemical synthesis. *Catalysts* **2018**, *8*, 223. [CrossRef]

17. Blánquez, A.; Rodríguez, J.; Brissos, V.; Mendes, S.; Martins, L.O.; Ball, A.S.; Arias, M.E.; Hernández, M. Decolorization and detoxification of textile dyes using a versatile *Streptomyces* laccase-natural mediator system. *Saudi J. Biol. Sci.* **2018**, *26*, 913–920. [CrossRef]
18. Rodríguez Couto, S.; Toca-Herrera, J.L. Lacasses in the textile industry. *Biotechnol. Mol. Biol. Rev.* **2006**, *1*, 115–120.
19. Legerska, B.; Chmelová, D.; Ondrejovič, M. Degradation of synthetic dyes by laccases—A mini-review. *Nova Biotechnol. Chim.* **2016**, *15*, 90–106. [CrossRef]
20. Ferrer-Miralles, N.; Domingo-Espín, J.; Corchero, J.L.; Vázquez, E.; Villaverde, A. Microbial factories for recombinant pharmaceuticals. *Microb. Cell Fact.* **2009**, *24*, 8–17. [CrossRef]
21. Couto, S.R.; Herrera, J.L.T. Industrial and biotechnological applications of laccases: A review. *Biotechnol. Adv.* **2006**, *24*, 500–513. [CrossRef]
22. Rodgers, C.J.; Blanford, C.F.; Giddens, S.R.; Skamnioti, P.; Armstrong, F.A.; Gurr, S.J. Designer laccases: A vogue for high-potential fungal enzymes? *Trends Biotechnol.* **2010**, *28*, 63–72. [CrossRef] [PubMed]
23. Fernandes, T.A.R.; Silveira, W.B.; Passos, F.M.L.; Zucchi, T.D. Laccases from actinobacteria—What we have and what to expect. *Postepy Mikrobiol.* **2014**, *4*, 285–296.
24. Guan, Z.B.; Shui, Y.; Song, C.M.; Zhang, N.; Cai, Y.J.; Liao, X.R. Efficient secretory production of CotA-laccase and its application in the decolorization and detoxification of industrial textile wastewater. *Environ. Sci. Pollut. Res.* **2015**, *22*, 9515–9523. [CrossRef] [PubMed]
25. Martins, L.O.; Soares, C.M.; Pereira, M.M.; Teixeira, M.; Costa, T.; Jones, G.H.; Henriques, A.O. Molecular and biochemical characterization of a highly stable bacterial laccase that occurs as a structural component of the *Bacillus subtilis* endospore coat. *J. Biol. Chem.* **2002**, *277*, 18849–18859. [CrossRef] [PubMed]
26. Ihssen, J.; Reiss, R.; Luchsinger, R.; Thöny-Meyer, L.; Richter, M. Biochemical properties and yields of diverse bacterial laccase-like multicopper oxidases expressed in Escherichia coli. *Sci. Rep.* **2015**, *5*, 10465. [CrossRef] [PubMed]
27. Sherif, M.; Waung, D.; Korbeci, B.; Mavisakalyan, V.; Flick, R.; Brown, G.; Abouzaid, M.; Yakunin, A.F.; Master, E.R. Biochemical studies of the multicopper oxidase (small laccase) from *Streptomyces coelicolor* using bioactive phytochemicals and site directed mutagenesis. *Microb. Biotechnol.* **2013**, *6*, 588–597. [CrossRef] [PubMed]
28. Arias, M.E.; Arenas, M.; Rodríguez, J.; Soliveri, J.; Ball, A.S.; Hernández, M. Kraft pulp biobleaching and mediated oxidation of a nonphenolic substrate by laccase from *Streptomyces cyaneus* CECT 3335. *Appl. Environ. Microbiol.* **2003**, *69*, 1953–1958. [CrossRef]
29. Kandasamy, S.; Devi, P.; Chendrayan Uthandi, S. Laccase producing *Streptomyces bikiniensis* CSC12 isolated from compost. *J. Microbiol. Biotechnol. Food. Sci.* **2017**, *6*, 794–798.
30. Eugenio, M.E.; Hrenández, M.; Moya, R.; Martínsampedro, R. Evaluation of a new laccase produced by *Streptomyces ipomoea* on biobleaching and ageing of kraft pulps. *BioResources* **2011**, *6*, 3231–3241.
31. Kuddus, M.; Joseph, B.; Ramteke, P.V. Production of laccase from newly isolated *Pseudomonas putida* and its application in bioremediation of synthetic dyes and industrial effluents. *Biocatal. Agric. Biotechnol.* **2013**, *2*, 333–338. [CrossRef]
32. Peter, J.K.; Vandana, P. Congo red dye decolourization by partially purified laccases from *Pseudomonas aeruginosa*. *Int. J. Curr. Microbiol. App. Sci.* **2014**, *3*, 105–115.
33. Arunkumar, T.; Anand, A.D.; Narendrakumar, G. Production and partial purification of laccase from *Pseudomonas aeruginosa* ADN04. *J. Pure Appl. Microbiol.* **2014**, *8*, 727–731.
34. Bugg, T.D.; Ahmad, M.; Hardiman, E.H.; Singh, R. The emerging role for bacteria in lignin degradation and bio-product formation. *Curr. Opin. Biotechnol.* **2011**, *3*, 394–400. [CrossRef] [PubMed]
35. Su, J.; Fu, J.; Wang, Q.; Silva, C.; Cavaco-Paulo, A. Laccase: A green catalyst for the biosynthesis of poly-phenols. *Crit. Rev. Biotechnol.* **2018**, *38*, 294–307. [CrossRef] [PubMed]
36. Sakurai, T.; Kataoka, K. Basic and applied features of multicopper oxidases, CueO, bilirubin oxidase, and laccase. *Chem. Rec.* **2007**, *7*, 220–229. [CrossRef] [PubMed]
37. Miñambres, B.; Martínez-Blanco, H.; Olivera, E.R.; García, B.; Díez, B.; Barredo, J.L.; Moreno, M.A.; Schleissner, C.; Salto, F.; Luengo, J.M. Molecular cloning and expression in different microbes of the DNA encoding *Pseudomonas putida* U phenylacetyl-CoA ligase. Use of this gene to improve the rate of benzylpenicillin biosynthesis in *Penicillium chrysogenum*. *J. Biol. Chem.* **1996**, *271*, 33531–33538. [CrossRef]

38. Sharma, P.; Goel, R.; Capalash, N. Bacterial laccases. *World J. Microbiol. Biotechnol.* **2007**, *23*, 823–832. [CrossRef]
39. Margot, J.; Bennati-Granier, C.; Maillard, J.; Blánquez, P.; Barry, D.A.; Holliger, C. Bacterial versus fungal laccase: Potential for micropollutant degradation. *AMB Express* **2013**, *3*, 63. [CrossRef]
40. Chojnacki, S.; Cowley, A.; Lee, J.; Foix, A.; Lopez, R. Programmatic access to bioinformatics tools from EMBL-EBI update: 2017. *Nucleic Acids Res.* **2017**, *45*, 550–553. [CrossRef]
41. McMahon, A.M.; Doyle, E.M.; Brooksa, S.; O'Connor, K.E. Biochemical characterization of the coexisting tyrosinase and laccase in the soil bacterium *Pseudomonas putida* F6. *Enzm. Microb. Technol.* **2007**, *40*, 1435–1441. [CrossRef]
42. Seeman, T. PROKKA: Rapid prokaryotic genome annotation. *Bioinformatics.* **2014**, *30*, 2068–2069. [CrossRef] [PubMed]
43. Marchler-Bauer, A.; Derbyshire, M.K.; Gonzales, N.R.; Lu, S.; Chitsaz, F.; Geer, L.Y.; Geer, R.C.; He, J.; Gwadz, M.; Hurwitz, D.I.; et al. CDD: NCBI's conserved domain database. *Nucleic Acids Res.* **2015**, *43*, 222–226. [CrossRef] [PubMed]
44. Marchler-Bauer, A.; Bo, Y.; Han, L.; He, J.; Lanczycki, C.J.; Lu, S.; Chitsaz, F.; Derbyshire, M.K.; Geer, R.C.; Gonzales, N.R.; et al. CDD/SPARCLE: Functional classification of proteins via subfamily domain architecture. *Nucleic Acid Res.* **2017**, *45*, 200–203. [CrossRef]
45. Santhanam, N.; Vivanco, J.M.; Decker, S.R.; Reardon, K.F. Expression of industrially relevant laccases: Prokaryotic style. *Trends. Biotechnol.* **2011**, *29*, 480–489. [CrossRef] [PubMed]
46. Tony, B.D.; Goyal, D.; Khanna, S. Decolorization of textile azo dyes by aerobic bacterial consortium. *Int. Biodeterior. Biodegrad.* **2009**, *63*, 462–469. [CrossRef]
47. Pereira, L.; Coelho, A.V.; Viegas, C.A.; Correia dos Santos, M.M.; Robalo, M.P.; Martins, L.O. Enzymatic biotransformation of the azo dye Sudan Orange G with bacterial CotA-laccase. *J. Biotechnol.* **2009**, *139*, 68–77. [CrossRef]
48. Mishra, V.K.; Sharma, H. Decolourization of textile azo dyes by *Bacillus* spp. In Proceedings of the International Conference on Emerging Trends in Traditional and Technical Textiles, Jalandhar, India, 11–12 April 2014.
49. Schlegel, H.G.; Kaltwasser, H.; Gottschalk, G. A submersion method for culture of hydrogen-oxidizing bacteria: Growth physiological studies. *Arch. Mikrobiol.* **1961**, *38*, 209–222. [CrossRef]
50. Bradford, M.M. A rapid and sensitive method for the quantitation of microgram quantities of protein utilizing the principle of protein-dye binding. *Anal. Biochem.* **1976**, *72*, 248–254. [CrossRef]

MDPI
St. Alban-Anlage 66
4052 Basel
Switzerland
Tel. +41 61 683 77 34
Fax +41 61 302 89 18
www.mdpi.com

*Catalysts* Editorial Office
E-mail: catalysts@mdpi.com
www.mdpi.com/journal/catalysts